WRITING FAST PROGRAMS:

A PRACTICAL GUIDE FOR SCIENTISTS AND ENGINEERS

WRITING FAST PROGRAMS:

A PRACTICAL GUIDE FOR SCIENTISTS AND ENGINEERS

John Riley

Viva Books

New Delhi | Mumbai | Chennai | Kolkata | Bangalore | Hyderabad | Kochi

First Indian Edition 2008

VIVA BOOKS PRIVATE LIMITED

- 4737/23 Ansari Road, Daryaganj, New Delhi 110 002
  E-mail: vivadelhi@vivagroupindia.net, Tel. 42242200
- Plot No. 76, Service Industries, Shirvane, Sector 1, Nerul, Navi Mumbai 400 706
  E-mail: vivamumbai@vivagroupindia.net, Tel. 27721273, 27721274
- Jamals Fazal Chambers, 26 Greams Road, Chennai 600 006
  E-mail: vivachennai@vivagroupindia.net, Tel. 28290304, 28294241
- B-103, Jindal Towers, 21/1A/3 Darga Road, Kolkata 700 017
  E-mail: vivakolkata@vivagroupindia.net, Tel. 22836381, 22816713
- 7, Sovereign Park Apartments, 56-58, K. R. Road, Basavanagudi, Bangalore 560 004
  E-mail: vivabangalore@vivagroupindia.net, Tel. 26607409, 26607410
- 101-102 Mughal Marc Apartments, 3-4-637 to 641, Narayanguda, Hyderabad 500 029
  E-mail: vivahyderabad@vivagroupindia.net, Tel. 27564481, 27564482
- Beevi Towers, First Floor, SRM Road, Kaloor, Kochi 682 018, Kerala
  E-mail: vivakochi@vivagroupindia.net, Tel: 0484-2403055, 2403056

www.vivagroupindia.com

Published by arrangement with
Cambridge International Science Publishing
7, Meadow Walk, Great Abington
Cambridge CB1 6AZ, UK

ISBN-10: 81-309-0839-5
ISBN-13: 978-81-309-0839-7

Published by Vinod Vasishtha for Viva Books Private Limited, 4737/23, Ansari Road, Daryaganj, New Delhi-110 002. Printed and bound by Anand Sons, Delhi - 110092.

To my daughter, Hunter, who helped develop this book between naps and learning to crawl

About the Author

John S. Riley began programming in the early 1980's and completed his first chemistry related software project in 1988: a custom microcomputer printer driver to print molecular orbitals on a dot matrix printer.

While earning his Ph.D. at the University of North Carolina - Chapel Hill, he began to explore quantum chemistry and Monte Carlo calculations to model experimental ion dissociation dynamics data in addition to writing low level code to collect that data. It was at this time that he began to think about the practical limitations to useful modeling on microcomputers.

In his chemistry career, Riley has conducted ion dynamics and photochemistry research, built and managed a forensic laboratory, taught college level Chemistry and Physics and is currently the owner and manager of DSB Scientific Consulting. In this capacity, he is involved in researching theoretical decomposition mechanisms of energetic materials and the numerical modeling of deflagration and detonation phenomena. His company provides computational chemistry and computational engineering services, including the development of highly optimized cross-platform numerical modeling applications.

PREFACE

This book is the culmination of much research in the quest to make a simple computational demo run fast enough to be useful. After many hours of reading and studying, and finally learning *something* about more efficient programming, I asked myself two questions: "Why are these techniques not universally employed in scientific and engineering codes?" and "Why is this information not in one place?"

In regard to the first question, I have, over the past twenty years, spent a fair amount of time working with my own small programs and the BASIC, C and ASSEMBLY source code of others. My comments in the book regarding inefficient practices in "real" code are most certainly *not* a criticism of those authors (how can I criticize those who did as I did?). Rather, in pointing these out in the text, I merely wished to show that slow techniques are used in real applications and there *is* room for improvement.

As for the second question, well, here it is. I don't pretend this book to be the end-all of scientific code optimization. Quite the contrary; I could easily list a dozen things not covered. Further, there are additional examples that could have been given. This book focuses on Windows and Linux based PC's, though the techniques are general, and can be applied to any platform. It was attempted to present the hardware specific material in a manner that is easily extended to other hardware.

I believe this book is a collection of practical tips, tricks and suggestions (Part II) with a firm foundation (Part I). I tried to pack as much useful information into this book as I could make relevant. In other words, I tried to write the book I needed five years ago.

Chapters 2 and 12 could be books in their own right. Therefore, with this material, I tried to hit the high spots. With these two chapters, the fine line between 'too long' and 'enough material to make sense' was quite elusive.

The demo results are presented for codes compiled with Visual Basic, Visual C++ (for Windows) and gcc for Linux. Some may question this choice of compilers (specifically, Visual Basic) for the demo results, but it was my hope to simply demonstrate that the techniques transcend language/compiler implementation and are general. In any case, I tried to provide demo versions to reach as wide an audience as possible.

Finally, it is my sincerest hope that you find this book useful in *your* code development efforts.

John Riley

FOREWORD

The relationship between computers and their scientific users has undergone several remarkable transformations. In the very early days (1950's), scientists interested in harnessing the power of the computer for their problems needed to be expert in programming at the bit level, thereby making computers accessible to only a few fringe scientists. However, with the advent of Fortran and BASIC compilers in the mid to late 1950's and 1960's, computers became user friendly permitting even those with limited skills to write programs to suit their needs. My own research has depended on computers ever since my graduate days in the late 1960's at Cornell University, where we had to punch IBM cards on monster machines and carry carefully ordered cards to the submission window. Data recorded on a tickertape by an IBM teletype machine permitted us for the first time to store and manipulate digital data. But, frequent trips to the computer center and long turn-around times were the inevitable part of the ritual. By the late 1970's laboratory computers by DEC (PDP 11/03) and floppy discs became available, which permitted us to finally integrate computers and experiments, and made possible easy manipulation and analysis of data. Each stage of computer development permitted us to speed-up and to improve the sophistication of our data analysis, thereby greatly enhancing the reliability of our results. It is, however, interesting that as fast as computers have advanced, the time required for data analysis has not been reduced. In other words, we have traded in computer speed for more sophisticated data analysis, or for the study of more complex chemical systems. The inevitable conclusion is that we can always use more computer power. Most of us have been satisfied with relying simply on increased computer speed to keep us at the forefront of our fields. However, computer power can also be enhanced by efficient programming, a concept that is quite unfamiliar to many of us. Efficient programming requires a more sophisticated understanding of computer architecture than many of us have acquired in our standard training as scientists. For many of us, the knowledge gap is sufficiently large to prevent us from making full use of the computer power that we currently have in our laboratories. This monograph by John S. Riley is therefore a most useful tool to help us bridge the gap. A few simple changes in computer codes can easily double the speed of the calculation, which the impatient will gladly accept, and others can trade in for longer integration times, or one more methyl group. Either way, we are better off.

Tomas Baer
Department of Chemistry
University of NC – Chapel Hill
Chapel Hill, NC

ACKNOWLEDGEMENTS

The author would like to thank the following for specific contributions that made this work possible. For ideas, reviews, conversation and keeping me honest in my presentation: David Sims, Tony Ellis and Harvey Tiffany (Carolina Computer Concepts); Ed Bradford (IBM); John Buchanan (Precision Crafted Ammunition); Tom Womack. For providing value comments and suggestions while reviewing the manuscript: Walter Murray, Kevin Morris (Carthage College), Tom Baer (UNC Chapel Hill) and Kurt Russell. For providing the Pentium III test platform for the Chapter 12 demos: Kurt Russell. For the inline pseudo random number generator algorithm: Ian Foster (Fermilab).

The following have provided material for the CD-Rom either knowingly by explicit permission or implicitly via the GPL. In either case, their work is greatly appreciated: Andrew Worsey Richards (CodePlay), Bill Gearhart (The ASP Emporium), Alavoor Vasudevan (PHP How-To), Stig Sæther Bakken and Egon Schmid (php_manual_en.chm file), Clark L. Coleman, Brendan Underwood and colin@nyx.net (Using Inline Assembly with gcc How-To), Kris Buytaert (OpenMosix How-To), Dave Jarvis (3-D Modeling How-To), Andr D. Balsa (Linux Benchmarking How-To), Jacek Radcjewski and Douglas Eadline (Beowulf How-To), Ram Samudrala (Linux Cluster How-To), Daniel Barlow (gcc How-To), Hank Dietz (Linux Parallel Processing How-To), David A. Wheeler (Program Library How-To), Manoj Warrier (Scientific Computing with Free GNU/Linux Software How-To), ASA, Inc. (the SciMath library info), Iain Nicholson (the libSIMD information).

For Help with Layout of Figure 6.1: John Holmberg (Visual Communication Productions).

Also, though we have never met or communicated in any way, I would like to thank programmer and author Andre LeMothe for inspiring me to be a better programmer in regard to run time performance.

Finally, for her complete support, encouragement and considerable patience, I would like to thank my wife, Rebecca C. H. Riley, M.D.

List of Figures

capable). The SSE routine is *slower* than the x87 and 3dNow! routine for *Par01* on both AMD and Intel CPU's. For *Par02*, the Intel implementation of SSE is superior to x87 code, but the AMD implementation of SSE runs slower than both x87 and 3dNow!. SSE shines for *Par03* on both AMD and Intel CPU's.

Figure 12.6: Basic procedure for coding parallel programs using socket objects. (a) After process creation, each worker process listens on the socket for data and instructions from the server process. (b) The server process sends data to slaves and waits for results.

List of Tables

CONTENTS

Chapter 1: Introduction to Code Optimization

In this Chapter, the motivation for optimizing code will be presented. In addition, a brief comparison of an easy-to-learn, rapid development time language (Basic) and a fast language (C) will be given. Finally, a general outline for the process of optimization will be discussed. This background information is presented before the Chapters on "Hardware" (Ch. 2) and "Operating Systems" (Ch. 3) in order to provide the context in which these system components will be explored.

1.1 MOTIVATION FOR WRITING HIGH PERFORMANCE CODE

When thinking of computers as scientific problem solving tools, we typically envision two distinctly different images. The first is the large mainframe "supercomputer" housed in some environmentally controlled special room and maintained by a full-time staff of dedicated systems experts. These systems, in the scientific context, are typically accessed via remote site terminals and are used by many practicing scientists to run existing codes. That is, there is relatively little day-to-day code development by the scientist, and the codes are run "as-is" to produce some computational result. Very often, these codes are used to mathematically model a physical system, such as the programs used for Computational Fluid Dynamics (CFD), Quantum Chemistry *ab initio* calculations, Statistical Mechanics models of reasonably large systems and searches on extremely large datasets.

In contrast, the desktop computer (alternately referred to as microcomputer or PC) is typically imagined to play more of a data acquisition and data analysis role. Further, these desktop computers may well be used as "smart terminals" for the mainframes mentioned above, and can employ software that acts as either input or output processors for the mainframe. In this type of configuration, the mainframe is used only for the most demanding part of a calculation, while the desktop computer is utilized for such tasks as pre-computing input values and graphing (and other visualization). One very important difference is the percentage of time the research scientist may spend developing software tools on the desktop. In a research environment, desktop software is continually developed to address this or that problem, and often such development is done on the fly with little planning and overall code organization.

However, recent advances in both PC hardware and software technologies have dramatically changed the role of desktop machines in scientific research. Increases in processor clock speeds, memory spaces, memory bandwidth, compiler technology as well as Component Object Model (COM) interfaces and Plug N Play peripheral devices have allowed the PC to be an even more easily integrated and useful scientific research tool. These advances, combined with lowering prices, contribute to PC's being used to solve hard core computational problems once believed solely accessible to large, parallel architectures. Indeed, the clustering of PC's into parallel architectures (such as the so-called Beowulf style cluster) is a growing field of microcomputer application and research.

So, in regard to small, affordable microcomputers, the question that consistently comes to mind is "What really can be done with a PC"? *Ab Initio* chemistry codes, Monte Carlo Statistics codes and 3-D fluid dynamics codes are a few examples of open source and commercially available packages running on PC's. With additional advances, the PC will undoubtedly continue to grow as a numerical modeling tool; as microcomputers expand their usefulness, so too the demands made upon them will continue to increase.

In addition, the purest may simply ask "just how fast can I get my task to run on this particular computer?" In this realm, speed improvements of 1% or even smaller can be considered milestones. Some 'speedfreaks' even revel in hours of programming to accomplish a less than 1% gain. While this approach may seem of limited practical importance, even small computational gains can accumulate in large, iterative procedures.

In short, the two questions outlined in the preceding two paragraphs provide the motivation for this book. With this motivation, the question "Just how do we generate faster code so that our computers can help solve larger, more complicate problems" will be addressed. As illustrated in the next section, many scientists writing computer code to solve practical problems are figuratively shooting themselves in the foot performance-wise, and are not taking advantage of the full potential of their computational machines.

To whet the appetite a little further, code examples will be presented in this book are demonstrated to run 1000's times faster (or more) than 'first try' coding of the algorithm. As an example, imagine something as simple as reducing a data curve fitting algorithm requiring five minutes to an execution time of less than 10 seconds.

1.2 SCIENTIST PROGRAMMER VS. COMPUTER PROGRAMMER

The discussion outlined in the preceding section begs a further question: "For whom is this book written?" To address this, a distinction is made between a "Scientist Programmer" and a "Computer Programmer." As Scientist Programmers, if we try to emulate the techniques of Computer Programmers, we will almost assuredly generate faster and more useful code.

There is an apparent difference between the programming styles of a scientist who knows a computer language and a computer programmer. Table 1.1 illustrates some key differences in the programming tendencies of these two classes of programmers. The table can be summarized by the following General Idea:

> *The Scientist Programmer tends to favor low development times in exchange for performance.*

If you find your programming style in the first category, the Scientist Programmer, this book will definitely introduce you to strategies to improve code performance. On the other hand, if you are in the second category, the 'Computer Programmer,' you may find in these pages some new tricks or revisit some forgotten ones.

To illustrate the characteristics of the Scientist Programmer with a practical example, consider a Monte Carlo procedure as might be used to model liquid-vapor equilibrium in a monoatomic system. This example, and others, will be developed throughout this book to illustrate the process of code optimization. To use this

Table 1.1 Comparison of the 'Scientist Programmer' to the 'COmputer Programmer'

Scientist Programmer	Computer Programmer
• Quick turnaround time from statement of programming problem to "finished code;" very short development time	• Researches/Tests for performance of code structures relevant to problem
• Relative lack of understanding of hardware	• Explores detailed knowledge of "likes and dislikes" of the hardware being programmed
• Compiler is a "black box" that turns source into executable	• Detailed knowledge of compiler and how different code structures will be compiled
• Quick design of data structures/little or no attention to memory management	• Detailed attention and careful planning of data structures and memory management
• Tends to use the first bug-free code as "finished code"	• Considers optimization an important part of the development process

procedure, the system energy is calculated as the sum of pairwise interaction energies of all the atoms in the system. Randomly selecting a large number of system configurations allows a large number of system energies to be calculated, and after weighting by a distribution function, the average system energy can be computed. By applying a theorem from Statistical Mechanics, the average system energy is equal to the observed energy if an 'infinite' number of systems is included in the average.

Mathematically, the process as described uses four basic equations. The first is the calculation of an average if the weights (or probabilities) and system energies are known:

Eq. (1.1)
$$<E> = \sum_{n=1}^{S} P_n E_n$$

with $<E>$ being the average energy, P_n is the probability (or weight) of system energy E_n and the sum is over S random systems (with S very large). The energies of the system are assumed, for the simple Monte Carlo procedure, to be distributed as a Boltzmann Distribution:

Eq. (1.2)
$$P_n = e^{-E_n/kT}$$

where k is the Boltzmann constant ($1.381 \cdot 10^{-23}$ J·K) and T is the absolute temperature at which the simulation is being run. The focus of our present attention, however, is in the computation of the actual system energies. The pair energy model is important in a variety of scientific and engineering problems so will serve as a prototype optimization calculation throughout this book.

The system energies are computed by summing the individual pair potentials over all the particles in the system, assuming an energy vs. distance function for the pair potentials. For our example, the Lennard-Jones potential can be used:

Eq. (1.3)
$$E_n = C \sum_{i=1}^{N} \sum_{j=1,\neq i}^{N} [(\sigma / r_{ij})^{12} - (\sigma / r_{ij})^{6}]$$

where the sums are over all particles in the system (N particles). In this function, C is the "strength" of the interparticle interaction (related to well depth), σ is separation at which the energy is zero (and is related to the position of maximum interaction) and r_{ij} is the distance between particle i and particle j. In two dimensions, this is

Eq. (1.4)
$$r_{ij} = \sqrt{(x_i - x_j)^2 + (y_i - y_j)^2}$$

where x and y are the Cartesian coordinates of particles i and j.

As a first attempt at coding this procedure, we could use the algorithm shown in Figure 1.1. This simple algorithm can be executed a large number of times so that the computed average energy is statistically meaningful. Ignoring for the moment the how the positions are selected, we will again focus on the energy calculation step.

To compute the system energies E_n, we might start with the code shown in **Listing 1.1**.

Listing 1.1 **Pure Scientist Style LJ Energy Computation in C.**

```
E=0;
for(i=1; i<=N; ++i){
     for(j=1; j<=N; ++j){
        // avoid division by zero when i=j
       if (i != j){
        E += pow(σ/sqrt(pow(x[i]-x[j],2) +
           pow(y[i]-y[j],2)),12) + pow(σ/sqrt(pow(x[i]-
 x[j],2) + pow(y[i]-y[j],2)),6);
       }
    }
}

E *= -C;
//avoids double counting the energies since i
//and j are looped over all N
E /= 2;
```

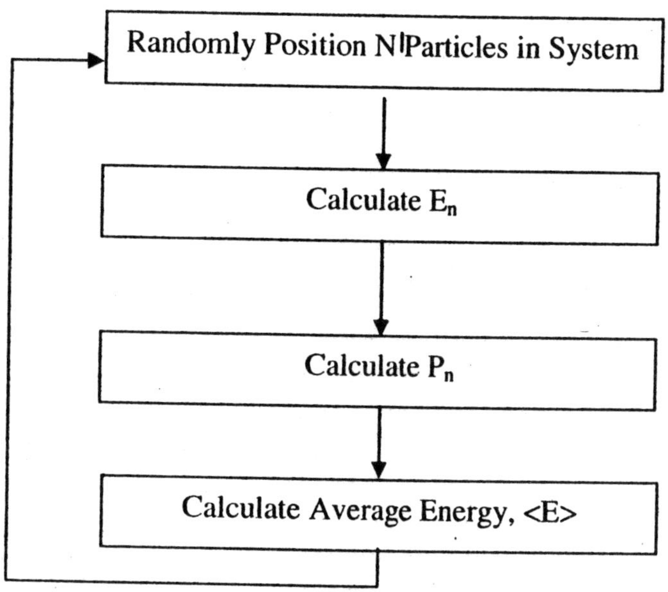

Figure 1.1. Flow chart for a simplified Monte Carlo procedure.

Here it is assumed that the array *x* holds the x positions of the particles, and the *y* array holds the y positions of the particles. While this code will certainly work, this algorithm is horribly inefficient and compiles to very slow code. However, to within a few minor changes, many Scientist Programmers would code the energy computation in this fashion.

The merit of this code structure to the Scientist Programmer is quite simple. This routine "looks like" the math at is it written on paper, say with Equation 1.4 substituted directly for r_{ij}. Therefore, one of the defining characteristics of Scientific Programming can be written:

> *The Scientist Programmer has the tendency to write code in a way that most resembles the mathematics as written in equations on paper.*

Recognizing the flaws in **Listing 1.1** is paramount to writing faster code, and quite a few optimizations will be made throughout this book.

A second major defining characteristic of the Scientist Programmer can be written:

> *The Scientist Programmer has the tendency to use fancy code structures.*

Or, as a corollary,

> *The Scientist Programmer has the tendency to write code suited for **SIZE OPTIMIZATION** rather than for **SPEED OPTIMIZATION**.*

As an example of this second defining characteristic, consider the two code listings for assigning values to a ten-element array shown in **Listing 1.2** and **Listing 1.3**.

Listing 1.2 **Short source code for array assignment**

```
//SomeValue is defined previously in code

//initialize n, the loop exit condition
n=10;

//do the loop
for(count=1; count<=n; ++count){

      R[Count] = SomeValue*count;

}
```

Listing 1.3 **Efficient, though longer, source code**

```
'SomeValue is defined previously in code
```

```
R[1]=SomeValue*1;
R[2]=SomeValue*2;
R[3]=SomeValue*3;
R[4]=SomeValue*4;
R[5]=SomeValue*5;
R[6]=SomeValue*6;
R[7]=SomeValue*7;
R[8]=SomeValue*8;
R[9]=SomeValue*9;
R[10]=SomeValue*10;
```

The Scientist Programmer would tend to favor **Listing 1.2** for four reasons:

1. Easier and quicker to type
2. Shorter code (fewer lines in source code and possibly smaller executable image)
3. Uses a "fancy trick" or complicated logic to perform a simple task.
4. The loop is general; the same code (in a function or library) can be used to assign arrays of various sizes.

As optimization strategies are developed throughout later Chapters, these two defining characteristics of the Scientist Programmer will be shown to generally lead to poor performance. By employing the styles of Computer Programmers, scientists developing applications will ultimately produce faster and more efficient code.

1.3 PROGRAMMING LANGUAGES

First, I am not going to advocate one language over any other in this section. Each computer language has its merits for the subset of software applications (or programmers) for which it was developed. Rather, the merits of a couple of languages will be considered to introduce the features to consider when comparing languages. A brief section on scripting/batch processing will also be given.

At the risk of angering all the Fortran (or other language) programmers, this brief outline of languages will cover only BASIC and C; most of the comments and techniques discussed in this book are applicable to any language, so sample code in only a couple of languages need be given. C and BASIC were chosen since they represent the two extremes (or near extremes) of pure performance on one end (C) and what we may call "ease of use" on the other (BASIC). It is believed that Fortran (or whatever is your favorite language) is within these two extremes, so for the sake of limiting redundancy, Fortran examples will not be given explicitly.

It should also be mentioned that fluency in any given language allows one to translate from another language to your language of choice. In general, each language can be coded to accomplish almost any task, so the question becomes "how easily" and "how fast." Due to its relative prominence in high performance computing, code examples in the text are given in most cases in C. Since BASIC represents a good pseudo language that is easily followed by programmers of many languages (who may not easily follow C), BASIC code for all the examples are included in Appendix E. In addition, Appendix E is included on the CD-Rom so readers may print them if they wish to follow along in BASIC (rather than C) to avoid having to constantly flip to the Appendix in the back. It is important to emphasize that the points being made throughout this book are general programming tips applicable to *any* language; I am not trying to teach coding specific algorithms in any particular language.

1.3.1 One Extreme: BASIC for Ease of Use

In my opinion, BASIC has four very important advantages that earn this language a place on desktop computers. First, the language is relatively easy to learn (and is, therefore, often a "first language"). Second, BASIC code has a natural language "look," an idea that relates to its appeal to beginners as well as to the first defining characteristic of the Scientist Programmer. In addition, BASIC has some very convenient string handling functions that make some 'bookkeeping' tasks in a program relatively easy. Finally, on a pragmatic note, BASIC remains a very popular language; many people program in BASIC so it is relatively easy to exchange programming ideas and find help when problems arise.

On the disadvantage side, BASIC requires very little of the programmer. While this may seem like an advantage (contributing to the ease of learning), in being "sloppy" tolerant, BASIC allows or encourages inefficient programming. Indeed, a fair portion of BASIC's reputation for being slow is due not to an intrinsic flaw in the language but rather to poor, undisciplined programming. Further, the BASIC language now has easily more than 200 keywords, contributing to source code compactness and ease of use. However, this is a tremendous vocabulary to master, and after 20 years of BASIC programming, I still have to refer to the Language Reference for syntax formats! Finally, BASIC is generally slower in execution than some other languages since the resulting compiled object code has the built-in generality of a language with over 200 words.

In weighing these pros and cons, I would argue that even with these drawbacks, BASIC continues as a useful language for small, quick utilities. For a beginning Scientist Programmer, quick code development allows fast development for algorithm testing purposes. Further, modern tools such as Visual Basic allow programmers to very rapidly develop User Interfaces for the Windows platform without extensive knowledge of objects and classes.

1.3.2 Another Extreme: C For Speed

On the pro side for C, little can be done to beat C's pure speed performance. There is a reason that much system programming (such as Unix and MS Windows) is done in C: these applications require the utmost in performance. Further, the hot-rod 3-D graphics, artificial intelligençe intensive programs (such as many modern games) are almost universally coded in C. There are several reasons for C's raw speed.

The ANSI Standard for C (introduced in 1983) established a language with only 32 keywords. With this small vocabulary, C can accomplish any programming task for which a logical algorithm can be written. By building up functions and procedures, C takes on some characteristics of a "high level language" such as BASIC, but the logical units are stripped for speed by coding for a particular application.

The small vocabulary also means that C is relatively easy to compile, and in source, much of the compilation is controlled. C is so clean, in fact, that in many cases properly coded C will compile to object code similar to that that would result from programming in ASSEMBLY Language. As a result, many procedures execute just as fast when coded in C as when programming in ASSEMBLY Language. Even when programming in ASSEMBLY does result in a performance gain, the gain is typically quite small for the general procedure. There are specific tasks, obviously, that are noticeably faster in ASSEMBLY.

On the con side, C requires more of the programmer than many languages and may therefore be harder for a beginner to learn than some 'slop tolerant' languages. C was from its very beginning a programmer's language. As such, C is not very tolerant of sloppy programming and its raw speed can easily be killed by *poorly designed code*. Therefore, to take full advantage of C's capabilities, it is paramount to properly design the code.

Technically, there is a difference between C and C++. C++ is an *extension* to the language and includes language level support for objects and classes, among other things (some people argue that C++ is a completely different language, but this is not true; virtually all of C is in C++, and any compiler for C++ will compile C). However, for the most part, the *numerical* routines on which this book is focused are coded in C (now commonly called 'straight C'). The notation C/C++ is somewhat cumbersome, so C will generally be used, with the possible exception being when it is important to emphasize C++ code or syntax.

MS Windows is intrinsically an object based Operating System, and C/C++ programming for Windows requires the programmer to understand data (and code) abstraction. Therefore, a beginning programmer writing C++ code for Windows must interact with C++ classes at least a little bit, and the learning curve can be quite steep. Development tools do flatten the learning curve somewhat.

1.3.3 Program Development Efficiency

In the final language comparison, however, an overall programming efficiency must be considered. Many beginner Scientist Programmers might prefer to learn one of the languages that is easier to learn. Development time must enter into the analysis of picking a language for producing an application that performs OKAY.

> *We will call the acceptable level of application performance the OKAY level...ie, a program that performs OKAY meets your needs and requirements for performance.*

If you can develop your OKAY application in two days using BASIC, for example, you will not benefit from programming C for ten days for an application that also performs OKAY (or better). That is, your overall performance takes into account development time, and it is important to include realistic performance goals when beginning a programming project. For example, a simple curve-fit of 1000 data points to two parameters will not tax modern computers; in this case, the language that is easier to learn or provides the quickest development time is the better choice. This book is dedicated to the idea of getting the peak performance for a particular language, no matter what language is selected. All of the examples given showed a performance increase for *both* of the two languages being considered, and these techniques are well documented to work in other languages (such as Java) as well.

Another issue that should be considered in the overall efficiency analysis is portability. If you are coding for one platform, this is not an issue. However, if you do plan to cross platforms or architectures with your code, portability is of prime importance. An example is using a desktop to develop and test a numerical subroutine on a desktop computer to 'upload' the working code to supercomputer when complete. C is generally more portable than BASIC (as is Fortran and, of course, Java and other more modern languages), but in any case, one must use care in employing platform specific functions or architecture specific optimizations. However, strictly speaking, programming with portability as a primary focus may hinder performance since platform specific optimizations may not be employed; this is explored in more detail in Chapters 2 and 3.

1.3.4 Scripting

Does scripting play a role in high performance computing? While it would be foolish to suggest writing a Lennard-Jones pair energy routine in MS Batch (assuming it can be done), script languages play an important part in overall development efficiency. For example, consider the following scenario as outlined by William Shotts, Jr on the web site http://www.linuxcommand.org. A network file server was crashing the network due to disk full errors. System programmers spent an entire working day producing C++ code to read the file system on the server and generate a catalog of file usage so the problem file(s) could be identified. However, another administrator achieved the same result, cataloging file use on the server, with a *single line* shell command. The power and usefulness of scripting should not be overlooked when considering efficiency.

Scripts are also quite useful when making first steps toward parallel computing. This is mentioned again in Chapter 12. Indeed, this can be extended to system administration in general. Even systems with robust Graphical User Interfaces, such as Windows, can be administered quickly and efficiently with scripts (and one might be able to bypass unwanted 'safeguards' in the system via scripts). Indeed, to quote Mr. Shotts, "... when you are a child, you use a computer by looking at the pictures. When you grow up, you learn to read and write."

1.3.5. Procedural vs. Object Oriented Languages

It may be obvious from the discussion on Section 1.3 that the design of a programming language itself in part determines performance. The way a language structures data, creates and uses subroutines, implements user interfaces, etc all impact the run-time performance of the code. Though languages such as C, BASIC, Fortran, Ada, Java and others are well known, there are many, many languages available. Some are highly specialized and others are for pure research in computer science. With so many languages, categorization schemes have been developed. Two of the most important categories for a programming language are whether the language is Procedural or Object Oriented.

Procedural languages arrange logical units in functions. Functions contain code and local data, and may receive data from the calling function. Function based languages appeal to programmers of mathematical code, since y=f(x) is coded as something like y=f(x). A function is a 'light' layer of abstraction in the program; the function name is simply the address of the code that performs the function. Function based languages tend to have less overhead, and the lighter abstraction can lead to source code that is easier to interpret. A key distinction is that a function lacks *encapsulation*.

Encapsulation is the inclusion of data with code that gives objects their character. An object is a much higher level of code *and data* abstraction; for example, a molecule object may contain data defining the number, kinds and positions of the atoms, charge of the molecule, spatial orientation as well as code the defines how to rotate and vibrate the molecule. The molecule object can then be manipulated with very short source code statements, such as molecule.rotate to rotate the molecule in space (note the 'dot' notation). One important thing to note about encapsulation is that the object may contain private data and functions, which add a great degree of robustness to the object paradigm (that is, users cannot change critical parts of the object; a classic example is account balance in banking software).

There are several recognized advantages to object oriented programming, especially at the large application level. These include modularity, reuse and encapsulation. Each of these favors the development and maintenance of the code, and does not address run-time performance. Due to the level of abstraction, objects incur considerable overhead on the system; the implementation of objects by the language and in source code by the programmer dramatically affects performance. Object oriented programming is one of those areas where just doing it and doing it right are two different things.

Programming with objects does have a key, important advantage (again, that favors the development of the program): a 'natural' abstraction of 'things.' It is very easy to create objects that match a physical problem, such as the molecule example given above. Provided one does not abuse functionality like multiple inheritance, deep levels of 'dot nesting' (that is, `object.method.method.method.method`) and making object methods overly terse (leading to many calls to many methods), object oriented programming is a powerful advance in computer software development.

1.4 GENERAL THOUGHTS ON OPTIMIZATION

Code optimization is not a single technique, but rather a software design philosophy. Producing high performance programs from codes written with little or no thought to optimization is difficult and inefficient. Therefore, it is recommended that all programs be written with the idea that optimization may follow, even if 'first try' performance is initially adequate. In this section, the process of optimization is briefly explored.

1.4.1 Why Optimize

The energy procedure in **Listing 1.1** will serve as a springboard to begin the discussion of optimization. One important software performance factor to be considered is *scalability*. Though several scalabilities will be considered in this book, the focus here is computational scalability:

> *Computational Scalability is the relationship between performance of a code and the size of the calculation (such as number of particles, dimensional mesh size, etc).*

If you curve-fit two parameters always to a fixed size array and your present code performs OKAY, computational scalability is not an issue. On the other hand, if you perform a molecular dynamics modeling, how the code performs as a function of number of atoms in the molecule contributes to limiting the size of problem that can be effectively addressed. As an example, we can use the Lennard Jones energy computation as shown in Figure 1.2 (and code in **Listing 1.1**) for N particles to examine the N dependence on the compute time. This procedure has two nested loops, a situation that occurs frequently in modeling of physical systems.

By letting t be the execution time of the actual energy calculation, the compute time

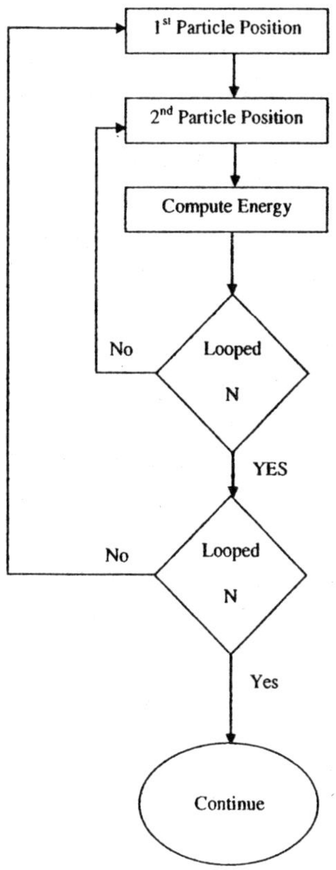

Figure 1.2. Flow chart for the pair energy computation algorithm. This procedure shows the nested loop structure of the basic algorithm. For *each* value of the outer loop, the inner loop must completely execute, giving the general n^2 dependence for n particles.

for the innermost loop (loop over the 2nd particle position) is clearly

Eq. (1.5) $$\text{Time, inner loop} = N \cdot t$$

Since the inner loop must be executed N times, once for each 1st particle, the time to compute the total system energy is

Eq. (1.6) $$\text{Time, total energy calculation} = N^2 \cdot t$$

To a first approximation, this leads to a general result, and can be stated as

> *The execution time for a nested loop scales as N*M*t, where N is the number of times through the outer loop, M is the number of times through the inner loop and t is the time to execute the code within the inner loop.*

Of course, we could generalize further to code that has code in the outer loop but not in the inner loop, and to additional levels of nesting. To summarize with an example, a 100 second (1 minute, 40 seconds) calculation using the procedure in Figure 1.2

will grow to a calculation requiring at least 400 seconds (6 minutes, 40 seconds) by merely doubling the number of particles.

Such N^2 scalability is clearly undesirable, and if possible, the algorithm should be modified to reduce the square dependence. Yet, even as performance hampering as such nested loops are, the Scientist Programmer (as defined in Table 1.1) continues to code them without questioning the inefficiency.

Three basic optimizations for nested loops are:

1. Reduce the exponential dependence (for example from N^2 to N dependence)
2. Reduce the loop sizes (for example, writing an algorithm using N/2 rather than N as the loop exit condition)
3. Reduce the time for execution of the code inside the innermost loop

A first simple optimization is readily apparent, and falls into the category of using the symmetry of the problem to reduce the sizes of the loops (improvement #2). In this example, it is noted that

Eq. (1.7) $$r_{ij} = r_{ji}$$

so that we are wasting computational effort by looping both inner and outer loops over *N*, then dividing *E*/2 outside the loops. To take advantage of this symmetry, the loops in **Listing 1.1** are recoded to those in **Listing 1.4**.

Listing 1.4 **Using symmetry to improve Listing 1.1**

```
for(i=1; i<=N-1; ++i){
      for(j=i+1; j<=N; ++j){
         // code to calculate pairwise energy}
         // goes here
      }
}
```

While there is still a nested loop structure (whose time is still given by N*M*t), algorithm simplicity has been traded for:

1. No more `if....else` decision structure to avoid division by zero
2. Outer loop is smaller by one and the total inner loop time is smaller by a factor of two.

The fundamental advantages of Improvement 1 will be considered in more detail in Chapter 2 and in Part 2.

Reducing the size of the inner loop (Improvement 2) will clearly show a benefit in terms of computational performance, but the scalability is not affected. However, there is a trade-off in that the code is rendered more obscure, though only slightly in this case. This leads to another general comment:

Optimization of code for speed will either

1. *make code (and likely the executable) larger, or*
2. *render algorithms more obscure, or*
3. *both*

There is clearly a trade-off between fast code and code that is small and easy to read.

In illustrating "Why Optimize" with the scalability, the first optimization to the system energy calculation was encountered. This basic procedure will continue to serve as an example throughout this text, and its optimization has only just begun.

On the subject of scalability, one of the most widely known and important numerical procedures enjoys the important success of improving computational scalability. The Fast Fourier Transform (FFT) improves the scalability of the 'traditional' Fourier Transform from an N^2 algorithm to an N log N one. The FFT can even achieve N scaling in certain special cases.

1.4.2 How Difficult is Optimization?

Of course, there is no single answer to this question. Some optimizations are typically simple, while others are far more complex. In any case, optimization "difficulty" is best measured in terms of development time: sloppy, inefficient code can be written quickly; robust, high performance code requires more planning and actual development time.The most practical approach is to ask, "How much optimization do I need?" Again, if your present code is serving with no complaints, improving it right now is a waste of effort. On the other hand, certain coding practices should be considered "normal" and are not really optimizations at all. "Style Optimizations" might be an appropriate name for these techniques. It is important to note that with sloppy tolerant languages, making these Style Optimizations a habit is more difficult.

Probably the "hardest" optimizations are those related to pure algorithm development. Picking the correct algorithm for the *specific* problem is of paramount importance. Writing of sort algorithms that scale as n^2 in *C: The Complete Reference*, Schildt states

> "When a [procedure] takes too long, it is usually the fault of the underlying algorithm. However, the first response is often, 'let's write it in ASSEMBLY code' ... If the underlying algorithm is inefficient ... the [procedure] will be slow no matter how optimal the coding ... The rule of thumb is that if the routine is not fast enough when written in C, it will not be fast enough in ASSEMBLY Language. The solution is to use a better ... algorithm."

Writing specifically about sorting algorithms, Schildt goes on to emphasize the key points mentioned in Section 1.4.1 for optimized code. Simple sorts have easy to follow source code, but are very inefficient at run-time. In contrast, faster, more efficient sort routines have source code that is considerably more difficult to interpret. Finally, the use of 'sentinels' in the data structure tremendously speed up sort routines, but destroys the generality of the routine. To emphasize this point further, in regard to search algorithms, Schildt states "... general purpose routines are sometimes too inefficient for use in demanding situations because of extra overhead created by their generalization." This last point is demonstrated for scientific applications in detail in Chapters 6, 7 and 8. If you don't want to expend a great deal of effort in actual algorithm development, part of the trick will be to utilize the latest algorithms and coding styles. An Internet search will often reveal websites with code examples of "advanced" algorithms.

Many Scientist Programmers use the Numerical Recipes series of books to develop computational codes. For what it is worth (and I neither endorse nor denigrate these books in saying this, this is provided simply for information), the algorithms contained therein are considered by many as drastically outdated, inefficient or even downright wrong. If you want the latest, "cleanest" algorithm for a particular task, a search of the computer programming and mathematics websites and literature is warranted. It should also be noted that there could be several years of time lag between algorithm advances and application of those advances by scientists. This last comment is particularly true of numerical procedures.

Optimization dealing with writing "compiler friendly" code can also be difficult to implement; it takes time, work and study to understand the compiler's algorithms. Knowing how your particular compiler interprets your source code may be called "advanced programming." Writing code that compiles and links correctly is only part of the process; writing code that compiles the way *you* want it compiled is a large step toward higher performance.

1.4.3 When and What to Optimize

With the exception of a little planning and attention to programming style, optimization is not typically attempted early in the development process. The key point here is to use a programming style that is open to optimization later should you find it necessary. For example, setting up a 10,000 line program with poorly planned data structures and memory management limits what you can do to improve performance. Another context in which "early optimization" applies are projects with multiple programmers providing code; these require special attention to details so that the interfacing of the various modules does not kill the performance. Throughout this book, however, we will assume a single programmer. To summarize, regardless of the project scope, writing code in good style will keep your optimization options open.

Generally, you will not want to begin true optimization at the beginning of the development cycle. The software development cycle may be represented as shown in Figure 1.3. A key step here is labeled "Profile Code," where the performance of the code is evaluated against objective measures. In particular, in many development projects, it is important to get an algorithm coded, debugged and operational to address the problem as soon as possible, *before* a serious attempt is made to improve overall performance. Only once the *need* to optimize is established should the time be invested in profiling and optimization.

Once it is decided that a performance gain is needed, profiling can begin. Profiling is the practice of evaluating code performance and pinpointing execution bottlenecks. If you plan to do a lot of optimization of large programs, there are software profilers available to assist in this process. For example, code profilers can count function calls, report time spent per function, identify inefficient code structures (such as multiple levels of nested function calls), etc, which can lessen the time spent profiling. Several code profilers are listed in Appendix A. Even without such software, however, modest sized codes can be profiled effectively manually.

One starting point for manual profiling is to have the program print execution times for major functions. This will help isolate execution bottlenecks. In addition, with some operating systems (such as Windows 2000 and Unix/Linux), one can access many features of system performance during execution such as %cpu time going to a process, disk cache read/writes (which may be important for very large data structures), memory allocation/deallocation and many more system statistics. These system level statistics can be logged over time to evaluate how a program is affecting overall system performance. It should be noted that so-called "home user" level operating systems do not provide the level of system evaluation required for serious profiling.

With objective measures of performance in hand (the execution times for major sections of code), actual optimization can begin. The initial focus is on those code structures that either

1. Most need optimization (are the biggest bottlenecks), or
2. Need optimization and are easiest to optimize

The first applies to code exhibiting one bottleneck that is much larger than the others and the second applies to code with several bottlenecks of nearly equal size.

By careful profiling, attempts to optimize code that is not degrading overall performance will be avoided. For example, it would do little good to optimize an array initialization that is only accessed once while leaving a 20x20 floating-point matrix inversion that is called iteratively thousands of times written inefficiently.

So, what are the steps to optimization? The flow chart in Figure 1.3 shows the generic (and non-descriptive) step 'optimize algorithm.' One approach to optimization is to 'blame the language' and translate to a faster computer language. Many Scientist Programmers assume that for improved performance, ASSEMBLY Language must be used. However, considerable improvement can often be made by reworking the slowest

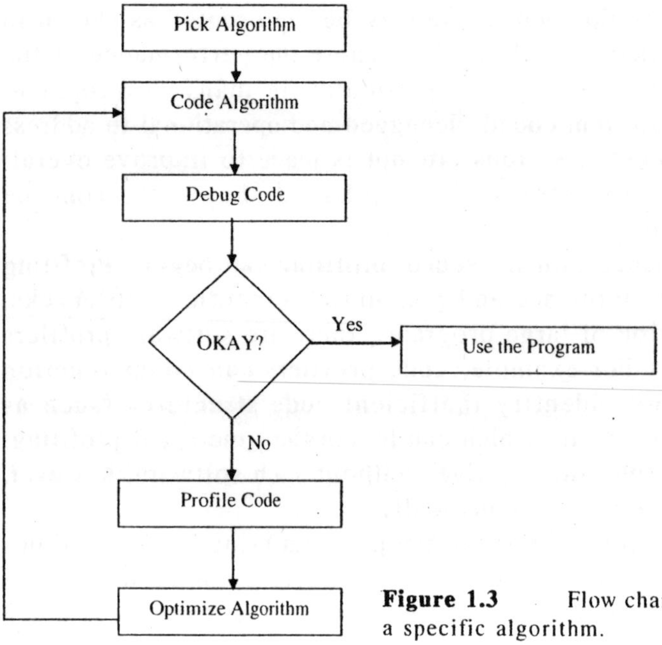

Figure 1.3 Flow chart of the process of code optimization, given a specific algorithm.

segments of the program. For example, one might ask "Is my data defined efficiently" and "Am I making unnecessary function calls or doing easy things the hard way?" The techniques presented in this book are a fair starting point, but they may be summed up in one simple statement: understand how the compiler turns your source code into object code for the platform you are running.

1.5 OVERVIEW OF SAMPLES PRESENTED IN THIS BOOK

It is important to note that the demo results presented in the book were real optimizations at the source code level. Compiler output listing files were checked to verify that the performance improvements observed were *not* the result of compiler optimizations or some trickery between source and executable image. In other words, the comparisons between two (or more) versions of the same algorithm were made as fair as possible, and the reader should observe similar improvements.

Because of their wide availability for desktop computers, x86 based PC's running Windows and/or Linux serve as the dominant test platforms for this book. Except as noted, there are three versions of the C source code for the demo programs. Two versions were written for Windows in C++ as Microsoft Foundation Class applications, and C for Windows Console applications; for the Windows versions, executables are included. In addition, a C version written to compile in Linux using gcc (and should therefore be usable with *any* Unix like OS using gcc) is included. Linux executables were not included since the gcc compiler allows architecture specific op-code scheduling via compiler command line switches. (Note that there are a couple of the Windows console and Linux console demos that are C++; these can be distinguished by the .cpp source file extension).

All of the Windows test results presented in this book were compiled with Visual Basic and Visual C++ 6.0, Service Pack 5, except where noted in the text. Visual C++ optimizations were not in general used for the simple repetitive examples due to compiler implementation errors (well documented, and briefly discussed in Chapter 4) in Visual C++'s global optimization code. The Athlon Linux test results were code compiled with gcc Version 3.2 and in general did employ compiler optimizations.

The sample programs include virtually no error or usability checking. These demos serve to illustrate coding styles and are not intended themselves to be 'stand-alone,' idiot-proof user-level applications. The author's in-house codes contain very little error checking since run-time performance is favored over being 'bullet-proof.' Input error checking via decision structures can be costly (as discussed in Chapter 2 and shown in Chapter 10), so the reader is encouraged to carefully weigh 'idiot-proofing' against performance. The demo programs are stripped to the algorithms and a basic user interface. In addition, the demo programs do not include measures for security. For example, input validity is not specifically checked.

The reader who favors object oriented (OO) programming may be quick to observe that OO styles are not typically used for the working parts of the code; straight function based programming was used. There are several reasons for this. First, the points being demonstrated are not in regard to overall program design; the working code is written similarly whether in a function based program or an object oriented one (that is, in an OO program, one must supply the member functions). Second, OO programming *can* add computational overhead to the program, especially if the objects are designed or organized poorly. Finally, not all Scientist Programmers are familiar with OO programming or compilers/languages that support objects. Therefore, to capture the interest and skill of a more moderate programmer, the simpler functional approach was used. A focus on OO style would limit this book as a resource.

The Monte Carlo procedure used as a model case study throughout the book was chosen since it contains quite a number of optimization points. The procedure as developed was intended as an educational tool: a Monte Carlo program for students to use in a computational chemistry lab exercise. As such, it had to be fast, since students do not seem to enjoy sitting and waiting ten minutes for a computer program to run. The first procedure written followed the example code given in Chapter 6 of *Introduction to Modern Statistical Mechanics* by David Chandler, but using a 'real' Lennard Jones pair potential rather than simplified nearest neighbors or hard disks. The code was unusable as an interactive demonstration on a 233 MHz Pentium computer (it required 20 minutes for a basic calculation; the final version, whose source code is listed in Appendix D, runs in less than one minute). Thus began the effort to optimize a PC thermodynamic Monte Carlo. Therefore, the program as outlined is not, and never was, specifically designed as a current, state-of-the-art

Statistical Thermodynamics Monte Carlo research tool; however, the programming techniques are certainly adaptable to research scale programs.

1.6 ADDITIONAL READING

Since this book is not a specific programming tutorial, but rather a brief guide for code optimization, several complimentary works are included here to aid the reader in finding resources more specifically geared to learning programming. This list is by no means exhaustive, but may serve as a useful starting point for further study.

Barton, John J. and Nackman, Lee R., *Scientific and Engineering C++: An Introduction with Advanced Techniques and Examples*, Addison Wesley Longman, Inc, 1994

Zachary, Joseph L., *Introduction to Scientific Programming: Computational Problem Solving in Mathematica and C*, Springer-Verlag New York, 1997

Ortega, James M., *Introduction to Fortran for Scientific Computing*, Oxford University Press, Inc, 1994

Berryhill, John R., *C++ Scientific Programming: Computational Recipes at a Higher Level*, John Wiley and Sons, Inc, 2001

Chapter 2: PC Hardware

The foundation for solid, high-performance programs is the hardware on which the program will run. It is true that C (and other languages) is quite portable; the source code need not be hardware 'aware.' However, ultimately it is machine code that executes, and understanding the basics of CPU architecture, instruction latency, addressing and basic ASSEMBLY Language structure provides the motivation for most of the optimizations discussed in Part II.

With many programmers, hardware portability is a goal. In this Chapter, the idea that generic, completely portable source code is not true high performance is introduced. For cases in which portability is important, conditional compilation may be a better approach than generic source code. For example, Demo *Hardware01* uses a defined macro to conditionally compile for Windows or Linux (this demo does *not* contain a hardware portability issue; this is included simply to demonstrate one method of including *specific* code in one source file). With conditional compilation, hardware optimizations can be utilized, but this does require considerable expertise for the programmer.

That some calculations have specific hardware requirements (rather than generalized, portable code) is illustrated nicely by the computers built for Lattice Quantum Chromodynamics calculations (*Physics Today* **57** (2004), 45). These computations require high local memory bandwidth with relatively small memory per processor. As such, these computers are often purpose-built. One such, the QCDOC, utilizes one IBM PowerPC processor with 4 megabytes of memory for each node.

This Chapter focuses on the x86 architecture popular in PC's. There is no way a single chapter in a single book can offer much depth in an overview of all possible hardware; rather, the goal is to introduce a common, affordable hardware along with the ideas important for effective use of higher level languages such as C, Fortran, BASIC, etc. Further, reference will be made to older CPU's since they form the foundation of modern processors and most compilers typically produce downwardly compatible code by default. Only CPU, bus and memory systems are considered.

2.1 BASIC SYSTEM ARCHITECTURE

At the most fundamental level, the CPU is composed of Complimentary Métal Oxide Semiconductor (CMOS) digital logic units such as flip-flops. Flip-flops are logic units that can be used to store state (like a 'memory transistor') by responding to an Input signal only if an Enable signal is 'on.' Further, flip-flops can be chained, so that as data pulses arrive, they propagate the chain. This arrangement would be a serial (versus parallel) register, where each data 'bit' passes from the low order individual flip-flops toward higher order ones. Such a serial flip-flop would require 8 pulses to populate data into an 8 bit register.

Of course, modern CPU's are composed of parallel registers, so that registers can be populated in one clock pulse. Other CMOS logic units exist in the CPU; the point is that the CPU consists of smaller logic units whose overall function is 'programmable' according to the state of logic units within the execution units. For example, the Arithmetic Logic Unit (ALU) may, during any given set of clock pulses, act as an adder, a multiplier, comparator, etc. The particular set of active logic units (or 'gates') is determined by the currently executing op-code (operation code) or machine instruction.

A parallel data path is termed a 'bus.' The CPU must have access to several key buses. One is the Address Bus, which allows a single Data Bus to access large pools of system memory. The other key bus is the Data Bus, which allows data to flow between the subsystems of the computer. Actually, there are two main data buses: Front Side Bus (FSB) and Back Side Bus. The Front Side Bus, sometimes called the System Bus, connects the CPU to system Random Access Memory (RAM). The Back Side Bus, on the other hand, connects the CPU to the Level 2 (L2) cache, which is temporary memory (either on the CPU chip or external to the CPU) that is much faster than main system RAM. Since many newer CPU's contain the L2 cache on-chip, the Back Side Bus would not appear on the mainboard (there are cases where the mainboard Back Side Bus would be used to access an L3 cache when L2 is on-chip).

Control of data on the Data Bus was historically provided by a variety of chips that performed specialized functions. On modern computer mainboards, larger scale integration of the data flow circuitry means that these circuits now appear in 'single chip' chipsets. There are two key components to the Data Bus chipset: Northbridge and Southbridge. The Northbridge controls data flow between the CPU, system RAM, Peripheral Component Interconnect (PCI) bus, L2 cache and Advanced Graphics Port (AGP) bus. On the other hand, the Southbridge 'sits on' the PCI bus, and controls data flow between CPU and input/output devices (such as serial ports, Universal Serial Bus (USB) ports, Integrated Drive Electronics (IDE) interfaces, etc.) Since the Southbridge resides on the PCI bus, data transfer between CPU and Southbridge is also influenced by the Northbridge.

It might be interesting to note that modern CPU's and the FSB likely do not run at the same clock speed. In old systems, such as the 80286, the CPU ran at *half* the system bus speed! Generally, the modern CPU is running at some integer *multiple* of the FSB clock. For example, the stock set-up for an Athlon (Thunderbird) 1200 based system has a 100 MHz FSB clock with a multiplier of 12. This means the FSB is running at 100 MHz (obviously), and the CPU itself runs at 1200 MHz. It theory, the 1200 MHz processor could also operate with a 133 MHz FSB clock and a multiplier of 9. This arrangement would give a greater data throughput to system memory while maintaining the processor at 1200 MHz. Such a system must support the higher FSB speed (both the Northbridge and physical RAM itself), and even then, the CPU may not be 'stable.' Such tweaking (termed overclocking) is risky to the hardware, but in some cases may produce a faster overall system.

The architecture of the mainboard, the chipset employed, bus clock speeds, etc, all influence the overall performance of a computer. It is common to compare benchmark data for two computers with the *same* CPU to find that one outperforms the other in terms of, say, Floating Point Operations per Second (FlOPS) acting on data in memory. The difference may be the flow of data between CPU and memory, which is controlled by the chipset and will be limited by bus frequency. It is very important to consider not only the CPU, but also the CPU within the entire system.

2.2 NUMERICAL REPRESENTATIONS

The representation of data in computer memory (and, similarly, other hardware storage units such as CPU registers) is important to consider in regard to performance. There is quite a difference between how the system manipulates the value 10 stored as an integer, an unsigned integer, a string of characters or in floating point format. For brevity, only numerical data (not character strings) is considered.

Integer data is rather simple to discuss. Each bit in the data represents a power of 2 in the base-2 representation of the number. Bitwise operators allow extremely fast manipulation of such data; for example, multiplication by 2 can be accomplished without a `MUL` op-code; instead, the `SHL` op-code can be used. Consider, for example, the decimal value 10, which in binary is

$$00001010$$

Shifting this binary data to the left produces

$$000010100$$

which is, of course, 20. Division by 2 could also be achieved by a right-shift. The x86 instruction set includes shifting instructions for integer and floating point data.

Addition of binary integer data is relatively straightforward, but how is subtraction achieved? The first step is to derive the *two's complement* of the data. The binary complement has the one's and zero's switched, so the binary complement of 10 is

11110101

The second step for subtraction is to *add one* to the two's complement. The binary representation is thus negated; the negative value can therefore be added. As a simple four-bit example, consider subtracting 3 from 5:

```
 0101     5d
-0011     3d
```

The two's complement of 0011 is 1100, which becomes 1101 when one is added. This value is added to 0101,

```
   0101                5d
+  1101               -3d
   0010,  carry 1           2d
```

and the carry is discarded. Signed integer operations are only slightly more complicated.

Floating-point representations require a minimum of two pieces of data: the mantissa and exponent (the sign is a third piece of data). That is, the single precision floating-point representation uses a binary scientific notation

Eq. (2.1) $$1.\mathrm{xxxxxxx} * 2^{\mathrm{Exponent}}$$

where xxxxxxx = the fractional portion of the mantissa. Single precision floating-point values in the IEEE 754 standard are represented in four bytes as shown in Figure 2.1 (double precision is similar, though the number of bits for both exponent and mantissa are greater). In this single precision format, there are 8 bits for the exponent, 23 bits for the *fractional portion* of the mantissa, and one bit for the sign. The actual exponent is offset by 127 so values less than one can be represented. As such, the exponent stored in memory is

Eq. (2.2) $$\text{Exponent Field} = \text{Actual Exponent} + 127$$

An interesting consequence of using binary arises in this notation. Since the single digit to the left of the decimal point is *always* 1, there is no need to store that in the number. This is why the least significant 23 bits store only the fractional portion of the value: the one is always implied.

Conversion between binary floating-point and decimal is straightforward and is similar to any other number base conversion. The bits for the mantissa represent $1/2^n$, so the base ten fractional portion of the mantissa is the summation

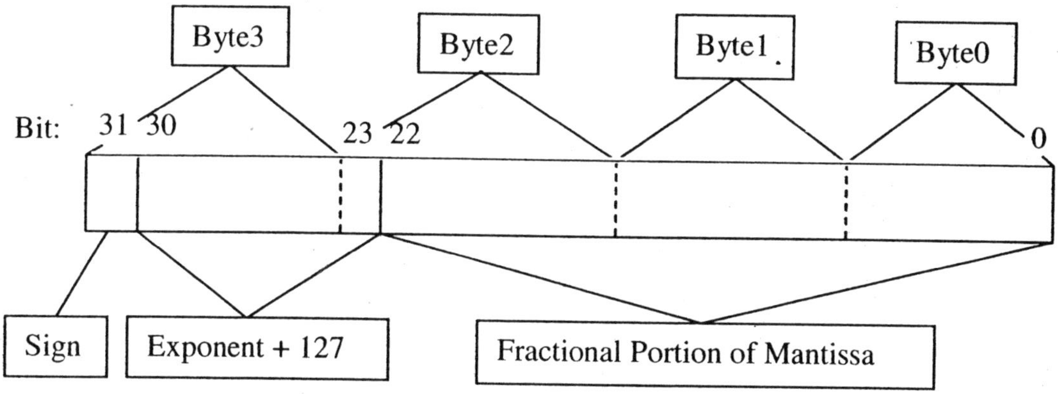

Figure 2.1 IEEE 754 format of a four byte, single precision floating point number. Note that the exponent field crosses the boundary of byte3 and byte2.

Eq. (2.3) $$FractionalMantissa = \sum_{n=1}^{23} a_n \frac{1}{2^n}$$

where a_n is the value of the nth bit (0 or 1). The smallest 'precision' for single precision values is therefore $\frac{1}{2}^{23}$ = 0.0000001. Conversion from decimal to binary floating point requires first the determination of an exponent, n, that yields a mantissa between 1 and 2 according to

Eq. (2.4) $$Mantissa = \frac{DecimalValue}{2^n}$$

Once the mantissa is determined, the *fractional* portion can be converted to binary using Equation 2.3. The values for $1/2^n$ up to n=23 (for single precision) are given in Table 2.1.

Table 2.2 lists several values in decimal and single precision floating-point representation. The reader is encouraged to practice these conversions by hand, in both directions, so that the manner of representing floating point data is understood. This is important for working with ASSEMBLY Language (such as reading ASSEMBLY output from a compiler) since immediate data are decimal or hexadecimal representations of binary floating-point values.

The convention shown above for representation of floating-point numbers in memory is a standard, but there is nothing magical about it. Indeed, until several years ago, Microsoft used a proprietary floating-point format, and the programmer sometimes needed to know which method was used in a given compiled program. This emphasizes a key point about floating-point representations. If, for some application, the IEEE standard does not suit your needs (perhaps you require additional bits in the exponent

Table 2.1 IEEE 754 bit mapping and decimal values for 2^x

IEEE 754 Bit Number	x	Decimal Value
22	-1	0.5000000
21	-2	0.2500000
20	-3	0.1250000
19	-4	0.0625000
18	-5	0.0312500
17	-6	0.0156250
16	-7	0.0078125
15	-8	0.0039062
14	-9	0.0019531
13	-10	0.0009766
12	-11	0.0004883
11	-12	0.0002441
10	-13	0.0001221
9	-14	0.0000610
8	-15	0.0000305
7	-16	0.0000153
6	-17	0.0000076
5	-18	0.0000038
4	-19	0.0000019
3	-20	0.0000010
2	-21	0.0000005
1	-22	0.0000002
0	-23	0.0000001

Table 2.2 Some real (non-integer) base ten values converted to IEEE 754 single precision floating-point format

Decimal Value	Sign Bit	Actual Exponent (base 10)	Exponent Field (base 2)	Mantissa (base 10)	Fractional Mantissa (base 2)	IEEE 754 Representation (base 16)
1.6780000	0	0	01111111	1.6780000	10101101100100010110100	3F D6 C8 B4
6.0000000	0	2	10000001	1.5000000	10000000000000000000000	40 C0 00 00
10.2300000	0	3	10000010	1.2787500	01000111010111000010100	41 23 AE 14
0.0087700	0	-7	01111000	1.1225600	00011111011000000001100	3C 0F B0 0C
-127.0000000	1	6	10000101	1.9843750	11111100000000000000000	C2 FE 00 00
-0.5000000	1	-1	01111110	1.0000000	00000000000000000000000	BF 00 00 00

and more precision in the mantissa, but need no sign bit), you can define your own representation. Note however, that this requires writing your own routines to handle mathematical operations, as the hardware instructions would not work properly.

The author encountered a similar situation in the late-1980's on a PDP-11 based system. The data was stored in a floating-point format and the software was not fully supported by the manufacturer. In short, data was collected from an instrument and stored on disk, but the analysis software was not very robust. By examining the floating-point format, the author was able to extract the needed data and customize the analysis routines in 'third party' software.

2.3 ADDRESSING

Each memory location in a given system must have an address by which it is 'known' to the CPU. In the early x86 systems, the addresses were segmented: the physical memory was divided into segments each containing 64 kilobytes, but segments could overlap. Individual addresses were specified by offsets from the segment. Using segmented addressing, it was possible to construct 20 bit addresses (1,048,576 bytes of memory) using only 16 bit registers.

Each address was constructed by adding an offset to the contents of a 16 bit Segment Register. 20 bit addresses were thus obtained by 'shifting left' the segment by four bits, then adding the 16 bit offset. For example, if the Code Segment register contains the value D671 H, a byte within the code segment with address DC831 H is obtained as follows:

D6710 H	CS register shifted left four bits
+ 6121 H	Offset, say contents of Instruction Pointer
DC831 H	Actual 'byte' address

Note that there is not a unique segment:offset combination for each physical memory address.

In a manner of speaking, one could say that such systems utilize 32 bit addressing, but this is misleading. It is true that the segment register was 16 bits and the offset is 16 bits, but they do not form the address by concatenation to Most Significant Word | Least Significant Word. That is, using the example from the previous paragraph, the address is not D6716121, which would be a 32 bit address. Since the addresses are formed by additive combination of 16 bit registers, segmented addressing is called 16 bit; 20 bit addressing is, however, more technically accurate.

Segmented memory gives rise to several important consequences to performance, most notably the memory 'model.' If all the addresses a program must access are within one segment, the model is SMALL and addressing is fast, since only the offsets need be computed or referenced. On the other hand, if the program must access addresses that span multiple segments, the segment register(s) must also be maintained. This constitutes additional 'work' for the CPU.

The segmented addresses just described were the only addresses accessible to pre-80286 processors. With the 286, protected mode addressing was added which allowed an address space far larger than the 1 MB 20 bit limitation. In protected mode, the segment registers in the 286 are indexes to entries in address tables called the Local Descriptor Table (memory space of the running process) and Global Descriptor Table (memory space accessible by ALL processes), both of which were 64 bits wide. Of the 64 bits in the table entries, only 24 bits were used in address construction. By adding the 16 bit offset to the 24 bit table entry, 24 bit virtual addresses were formed.

Newer x86 class CPU's utilize 32 bit registers. Therefore, in 32 bit systems, the memory space is mapped 'flat' as linear addresses 0 – (2^{32}-1); there are 4,294,967,296 available memory addresses in a 32 bit flat system. When an operating system is referenced as 16 bit or 32 bit, the addressing mode is being specified: Win32 (Windows 98 and later) utilizes the 32 bit flat addressing exclusively.

The CPU can access data by having the address specified directly, such as the `MOV AX,DS:[1234]` instruction. (The colon is an Intel format ASSEMBLY Language operator used when the segment is specified). This instruction loads the *value* stored at address DS register:1234 into the AX register. Note that this syntax is different from `MOV AX, 1234`, which would put the value 1234 into the AX register. The latter is called *immediate* addressing (it uses immediate data explicit in the instruction), while the former is *direct* addressing.

Another addressing mode that allows the CPU registers to be used somewhat like pointers is indirect addressing. In this mode, CPU registers point to addresses; offsets from these pointers can be given as either indexes or explicit offsets. In this way, data structures such as arrays and tables can be referenced in physical memory, much like using index variables in a high-level language. For example, suppose the EBX register points to byte array element A(0); The value of the EDI register can then be used to reference the different elements as an index (or, in this example, the offset could be given explicitly). For example, `MOV AX,[EBX][EDI]` moves the EDIth value into AX (note the absence of the colon separating the two register references). The array does not have to be single byte data. For example, if the array is single precision floating-point data, each value is 4 bytes. Therefore, the EDI register holds a multiple of 4.

It should also be mentioned that in ASSEMBLY Language, addresses can be given symbolic names (like a function name). Symbols can also be assigned to the *value* at an address (like a variable name). It is therefore important to recognize that an ASSEMBLY instruction such as `MOV EAX, MySymbol` can have different meanings depending on how the symbol `MySymbol` is defined. If `MySymbol` is a 'variable,' its value is used (indirect addressing); if `MySymbol` represents an address label, it is used as an immediate.

2.4 BASIC CPU ARCHITECTURE AND INTRODUCTION TO ASSEMBLY LANGUAGE

As one can imagine, a CPU is a complex system of smaller functional units. The details of these units and *how* they function are well beyond the scope of this book, but an overview is useful for providing the basis for optimization. In particular, the CPU is comprised of control units, data management units, arithmetic logic units for integer math (ALU), floating-point units, etc. The basic instructions that control the function of these logic units form the "Instruction Set Architecture" (ISA, or simply "instruction set") of the CPU. The binary (or other numerical) representation of each instruction is termed 'op-code,' and the symbolic mnemonic representations are the ASSEMBLY Language Instructions. For example, the x86 ASSEMBLY mnemonic `PUSH AX` (push contents of the AX register onto the stack) assembles to op-code 50 H (or 80 in base ten). ASSEMBLY Language also consists of syntax rules, assembler directives and symbolic definitions that ease the construction of op-code images.

The ASSEMBLY Language examples given in this Chapter, as well as the book in general, follow the Intel instruction format: `operation destination, source`. This convention is used not only in Intel based Assemblers, but also for the 'inline Assembly' used in Microsoft Visual C++ (which is how Windows ASSEMBLY is accessed in this book). Please note, however, that this differs from the default AT&T convention that is used for inline Assembly in gcc, the C compiler used commonly used for Linux. Therefore, for Linux examples that utilize inline ASSEMBLY, the instructions follow the AT&T format of

```
operation source, destination
```

so the operands 'appear' backward (when compared to the examples in the text). The ordering of source and destination operands are not the only differences between the Intel and AT&T instruction formats. The syntax used for addressing (including how symbolic and immediate data are referenced) differs as well. ASSEMBLY output from the gcc compiler can be written in the Intel format with the `-masm=intel` switch. Further, for Inline ASSEMBLY, gcc not only follows the AT&T instruction format, but also uses 'Extended Inline ASSEMBLY,' a powerful syntax for ASSEMBLY Language. Extended ASSEMBLY is very efficient in regard to optimization but is generally less readable for those new to ASSEMBLY Language. For an excellent beginner tutorial on gcc Inline ASSEMBLY (that includes comparison's of the Intel and AT&T instruction formats), see "Using gcc Inline ASSEMBLY" on the CD-Rom.

CPU control is provided by a number of specialized units. These include circuitry to fetch, decode, schedule, execute and retire instructions, as well as writing results, transferring data, etc. The control units are not typically under direct control of the programmer, but 'feeding' code easy for these units to 'digest' constitutes key optimizations when using high-level languages.

The operation of the CPU is ultimately controlled by the system clock. Each CPU clock pulse provides an opportunity for signals in the CPU to change state. Each clock pulse represents a "cycle," so that clock cycles are the fundamental unit of time within the CPU. Each CPU instruction requires that the CPU can execute requires a certain number of cycles to execute (technically called *latency*), and the timing during the execution cycle can be quite complex. Section 2.5 summarizes code execution and timing in more detail.

The data used by the CPU can be stored in several possible places. Data access speed for the CPU can be visualized as being related to 'distance.' The fastest data access is to CPU *registers* which are *in* the CPU and therefore closest. Other locations in increasing distance include L1 cache, L2 cache, system memory, local disk and network. The registers are extremely fast internal memory units and are typically used for temporary storage. Registers are categorized according to typical use, and in pre-386 processors, some could only be used for certain purposes. Sample uses include General Purpose (available to the programmer for math operands, for example), Index and Segment (memory addressing), Flags (CPU state) and Control. The Flag and Control registers cannot generally be set programmatically, though there are instructions to set or change certain flags. A functional diagram of 80286 registers (for simplicity) is given in Figure 2.2.

It is important to note that even on 8 or 16 bit CPU's, individual registers can sometimes be paired to operate as larger registers. For example, the pairing of the AL and AH 8088 registers to form the AX register allowed the CPU to operate on 16 bit data. 80386 and newer CPU's utilize 32 bit registers such as EAX (symbolizing the Extended AX register).

The General Purpose registers hold data or receive output from basic operations, such as those performed by the integer ALU. For example, the x86 instruction set includes an op-code for `ADD AL, AH`. This instruction, in ASSEMBLY Language Mnemonics, adds the data in the AH register to that in the AL register. The result is stored in AL, so that the AL operand is lost. Data in registers can also be added to immediate data (such as `ADD AL, 6`) or values stored in main system memory (`ADD AL, memaddress`).

Table 2.3 shows the latency for several simple instructions on several CPU's. In this table, the cycles given are for actual execution; these times do not include fetch, decode or other pre-execution tasks performed by the CPU. Note that adding to memory directly is relatively costly (as memory transfers are in general) in terms of op-code execution time, but adding to memory data eliminates several transfer steps. The following example illustrates how the latency data can be used to optimize

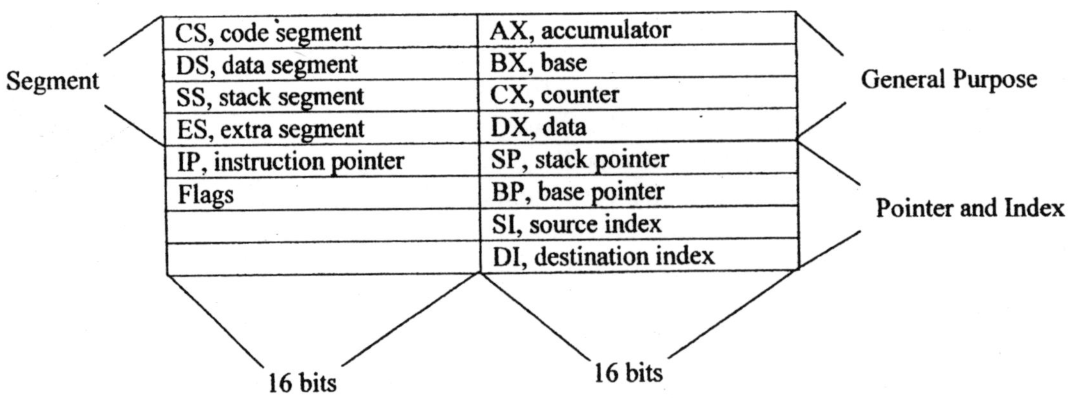

Figure 2.2 Functional diagram of the 16 bit 80286 registers.

Table 2.3: Some x86 instruction latencies for 286, 486 and Pentium processors

Instruction[a]	286	486	Pentium
ADD mem, reg	7	3	3
ADD reg, imm	3	1	1
ADD reg, reg	2	1	1
ADD reg, mem	7	2	2
CMP reg, imm	3	1	1
DEC reg	2	1	1
INC reg	2	1	1
JNZ	3/7+m[b]	1/3[b]	1
JMP	7+m	3	1
LEA reg, mem	3	1-2	1
MOV mem, reg	3	1	1
MOV reg, imm	2	1	1
MOV reg, mem	5	1	1
MOV mem, imm	3	1	1
MUL reg	21	13-26	11
POP reg	5	1	1
POPA	19	9	5
PUSH	3	1	1
PUSHA	17	11	5
XOR	2	1	1

a. imm = immediate data, mem=memory reference, reg=register reference, m=number of bytes in the next instruction to execute.

b. two values are given for conditional jumps: latency for no-jump/latency for jump

ASSEMBLY Language code.

There are at least two ways two unsigned 16 bit integer operands stored in memory can be added on the x86 based system, as shown in **Listing 2.1** and **Listing 2.2**. Suppose that the two pieces of data are stored at memory locations `memloc1` and `memloc2`. To keep the comparison fair, the result is stored back into `memloc2`.

Listing 2.1 x86 ASSEMBLY Language listing for 16 bit integer addition

```
MOV   AX,  memloc1      ; put value at memloc1 and
                        ; memloc1 + 1
                        ; into AX register
MOV   BX,  memloc2      ; put value at memloc2 and
                        ; memloc2 + 1
                        ; into BX register
ADD   AX,  BX           ; add data, result goes into
                        ; AX
MOV   memloc2, AX       ; store result back into
```

```
                         ; memloc2 and
                         ; memloc2 + 1
```

Listing 2.2 Alternate x86 ASSEMBLY Language listing for 16 bit integer addition

```
MOV  AX, memloc1     ; puts 2 bytes starting at
                     ; memloc1 into AX
ADD  memloc2, AX     ; add 2 byte data in AX to 2
                     ; byte data stored
                     ; at memloc2 and memloc2 + 1
```

Table 2.4 shows the time, in CPU clock pulses, required to execute these two nearly equivalent code structures. They are nearly equivalent since the INPUT states are equal and though OUTPUT states are not equal (in **Listing 2.1**, the BX register is altered without storing its initial value), the math result is stored in memloc2. If we assume that the initial value in BX is *not* needed, a total latency comparison of the procedures as written is fair. To contrast the two approaches, note that **Listing 2.1** requires the use of an extra General Purpose register and consists of more numerous, though faster, individual op-codes. On the other hand, if BX must be preserved, the temporary storage retrieval of BX (on the stack) will increase the data transfers and hence the latency of **Listing 2.1**.

This simple example illustrates several key points. First, the new CPU's (486 and Pentium) outperform the 286 even when absolute clock speed is not considered. Recall that these cycles refer to the number of (CPU) clock pulses. Second, how the code is compiled (suppose the two ASSEMBLY listings in **Listing 2.1** and **Listing 2.2** are compiler outputs from your high level language of choice) may impact the execution time. Code compiled to **Listing 2.1** (in this overly simple example) is slower than code compiled to **Listing 2.2** when both are run on a 286. Third, optimization may be hardware specific; in this example, a 286 optimization favoring **Listing 2.2** over

Table 2.4: 286, 486 and Pentium latencies (in absolute CPU clock cycles) for **Listings 2.1-2.6** and **2.8**

Code	286	486	Pentium
Listing 2.1	15	4	4
Listing 2.2	12	4	4
Listing 2.3	20	6	6
Listing 2.4	35	11	9
Listing 2.5	31	28[a]	13
Listing 2.6	281	85	16
Listing 2.8	56	18	12

a. assuming worst case latency; with best case, 15.

Listing 2.1 is lost on the new CPU's (assuming the initial value in BX is not needed). Finally, the newer CPU's are fairly optimized for basic instructions.

Of course, the situation changes a little if both operands need be preserved, as well as storing the result. In this case, the only change made to **Listing 2.1** is the final line:

```
MOV  memloc3, AX
```

used to store the result to a third memory location. This minor change does not change the execution time: 4 cycles are needed on the Pentium. However, for **Listing 2.2**, more work must be done, and it must be remembered that there is no instruction for `MOV mem, mem`. One such solution is shown in **Listing 2.3**.

Listing 2.3 Variant of Listing 2.2 that preserves both operands

```
MOV  AX, memloc1      ; put word at memloc1 into
                      ; AX
MOV  memloc3, AX      ; put word in AX into
                      ; memloc3, these two
                      ; instructions copy data
                      ; from memloc1 to memloc3
MOV  AX, memloc2      ; get the second operand
                      ; into AX
ADD memloc3, AX       ; add second operand to
                      ; the first, already
                      ; stored in memloc3
```

Using the data in Table 2.3, one can readily see that **Listing 2.3** requires 6 cycles on the 486 or Pentium (or 20 for the 286!). The added complexity of addition with preservation of *both* operands, the choice of adding to memory versus using two General Purpose registers becomes relevant. Clearly, very sophisticated logic would be required in the compiler to decide between these cases (both operands need be saved or not, initial BX preserved or not). The real question is "What is your *compiler* actually *doing*?"

This simple integer addition example shows that one can, to some degree, engineer code for optimal performance at the op-code level. Examining how a particular high-level language is compiled to ASSEMBLY instructions teaches the programmer much about optimizing algorithms and code even at the higher level. Throughout Part II, compiled code examples are given, so the reader should have the basic ASSEMBLY Language skills discussed in this Chapter.

General Purpose registers can be used for a variety of functions, such as loop counters (typically CX or ECX), comparison operands for branch testing, etc. Data pointers, on the other hand, were historically held in special purpose Index Registers. However, beginning with the 386, General Purpose registers *can* be used as index registers. One important index register is the *stack pointer* that points to the current 'top of stack.' The stack is a "Last In – First Out" (or LIFO) data structure in memory that the CPU uses for temporary storage. When data is `PUSH`ed onto the stack, the SP register is incremented so the next data is `PUSH`ed to the next memory location. Stack instructions include `PUSH` and `POP` (and several variants), and manipulation of the stack is important when functions and procedures are called (Chapter 6).

The Flag register supplies information about CPU state to the program. Operations can set flags, and these flags can be used for testing. Figure 2.3 lists the flags available in the Flag Register, again using the 286 for simplicity. For example, an addition may cause a 'carry,' such as a 32 bit addition producing a result larger than 32 bits. The `ADC` instruction allows the carry bit in the Flag register to be set in such a circumstance. By testing for the carry flag following an ADC instruction, programs

Bit	15	14	13	12	11	10	9	8	7	6	5	4	3	2	1	0
Flag		NT	IO	PL	O	D	I	T	S	Z		A		P		C

Flag	Brief Description
NT	Nested Flag (Protected Mode Only), set if task called by JMP
IOPL	I/O Privilege Level (Protected Mode Only)
O	Overflow Flag, set on numeric overflow
D	Direction Flag, direction of string manipulations
I	Interrupt Enable Flag
T	Trap Flag, puts cpu in single step mode for debugging
S	Sign Flag, set if operation produces a negative result
Z	Zero Flag, set if operation produces a zero result
A	Auxiliary Carry Flag, set if carry out or 4th bit
P	Parity Flag, set if result has 'even' number of 1's
C	Carry Flag, set if result carries out of most significant bit

Figure 2.3 Bit mapping of the 80286 Flag register.

can use this to emulate larger bit-space integer math. Further, a program may need to branch if the Carry Flag is set to implement such emulation code. Another useful flag is the Zero Flag which is set if an instruction results in a register being 'zero'd.' This is used to perform loop operations in ASSEMBLY Language. For example, **Listing 2.4** performs the integer multiplication `6*2` (pretending for a moment the x86 instruction set does not include a `MUL` instruction!).

Listing 2.4 x86 ASSEMBLY listing for integer multiply using repeated ADD

```
        XOR   AX, AX          ; common way to zero
                              ; a register
        MOV   CX, memloc1     ; CX is "counter"
                              ; register, loop 2
                              ; times; memloc1
                              ; contains 2
mulloop:
        ADD   AX, memloc2     ; address of this
                              ; instruction is
                              ; "mulloop"
                              ; memloc2 holds 6
        DEC   CX              ; decrement the CX
                              ; register
                              ; Z flag set if
                              ; DEC CX results in 0
        JNZ   mulloop         ; jump on NO ZERO (in
                              ; flag register) to
                              ; address mulloop
```

Careful examination of **Listing 2.4** shows that when CX=0 and the Zero Flag is set, AX contains $12 = 6 \cdot 2$. When CX = 0, the "Jump No Zero" `JNZ` instruction "fails" and flow continues to the next instruction; while CX > 0, the code jumps back to the address `mulloop`.

How does this jumping occur? The CPU contains a number of special Control Registers such as the Instruction Pointer or IP. There is no instruction to set the IP directly (i.e., one cannot `MOV IP, address`), but the IP can be set 'manually' using an unconditional `JMP` instruction. As one may guess, the IP contains the address of the next instruction to execute. The conditional jump `JNZ` (and similarly for its cousins in the instruction set) causes the CPU to test the zero flag. The IP is then set for the next instruction after the `JNZ` or loads the IP with the address symbolized in the example by `mulloop`. Recall that `mulloop` is the address of the first byte of the three-byte op-code for `ADD AX, 6` (byte one is the op-code for "add 16 bit immediate to AX and the second and third bytes are the 16 bit immediate data; 6 in this case).

The IP incrementing to the next instruction is generally faster than forcing an address into IP. This is why two latencies are given for conditional jump instructions: the smaller value is for when the jump is not made and the larger value is for when the jump is made. Data for **Listing 2.4** (excluding the `XOR` instruction) are also given in Table 2.4. For the 286, the multiplication requires 35 cycles, the 486 requires 11 cycles and the Pentium requires 9 cycles. As in the simple memory access and integer addition cases, hardware evolution has reduced the cost of conditional jumping, but it should be noted that programs that have many branches constantly must modify the IP 'artificially' and may have additional performance consequences.

The instruction set includes a MUL instruction (with one factor being in AX) for unsigned integer multiplication. This begs the question: "Is MUL faster than simple repetitive looping?" The answer depends on the factors multiplied. When the multiplication is performed on 16 bit operands (as the example in **Listing 2.1**), the latency for the MUL instruction is given in Table 2.3. So, for 6 · 2, **Listing 2.4** becomes

Listing 2.5 x86 ASSEMBLY listing for integer multiply using MUL instruction

```
MOV   AX,  memloc1      ; operand 2 is stored at
                        ; memloc1
MOV   BX,  memloc2      ; operand 6 is stored at
                        ; memloc2
MUL   BX                ; multiplies AX · BX, the
                        ; AX register is implied
```

The latency for this code is also shown in Table 2.4. In this particular example, the looped addition outperforms the MUL instruction! However, it is clear to see that for larger operands, using the MUL instruction is the faster code. Again, the key question is 'how does the compiler compile the source code?' It is probably safe to say that most compilers will compile 6*2 in a high level language to code similar to **Listing 2.5**, and clearly this not the best choice for these factors. In the general case, however, the MUL instruction is favored, since there are far more combinations of unsigned integer factors for which the multiplication is faster using MUL.

Actually, there is a subtle, yet very fast, way to optimize simple unsigned integer multiplications (especially if both factors are relatively small); this technique might be useful to multiply an extremely large integer array by a constant such as for scaling a matrix. The technique uses the fact that the *address offset* of a memory location is given by the equation

Eq. (2.5) $$Address = Immediate + Base\_Pointer + (Index\_Pointer \cdot Index\_Scale)$$

which is returned by the LEA (Load Effective Address) x86 instruction. Examination of Equation 2.5 shows that two integers m and n can be multiplied if: $Immediate = 0$, $Base\_Pointer = Index\_Pointer = m$ and $Index\_Scale = n - 1$. For example, to multiply 25 x 9, the single line

```
LEA eax, [25 + 25 · 8]
```

leaves 225 in eax. The interesting thing to note here is that since the LEA instruction is used in addressing, it has been engineered to execute in *one cycle* on Pentium class processors, making this method *ten times faster* than using `MUL`. Since this is an addressing instruction, and valid data types are 1, 2, 4 or 8 bytes, the valid values for *Index Scale* are 1, 2, 4 and 8. Therefore, using this technique means arbitrary integers can be multiplied by 2, 3, 5 and 9. If your compiler produces `MUL` for these simple integer multiplications, you may consider the use of inline ASSEMBLY in time critical applications. It should be noted that the gcc standard C library libc defines macros that use this technique for `times9`, `times5` and `times3`. There are other, similar, shortcuts for very fast integer multiplication.

The Floating Point Unit (FPU) operates a little differently than the integer ALU. For floating point arithmetic, the General Purpose registers are not used (partly because they are not large enough to hold 64 bit double precision data). Operands for the FPU are placed in *numeric registers*. For example, the 287 had eight numeric registers which could be used as 'normal' registers or could be used as a stack; these registers were labeled 0 – 7. When used as a stack, the symbol ST (Stack Top) is used to denote the register *currently* at the top of the stack, which is also ST(0). For example, Numeric Register 3 might be the top of the stack, so that ST(0) = ST = 3 (in this example, then, ST(1) = Numeric Register 4, ST(2) = Numeric Register 5, etc). These registers hold Floating Point, Integer and BCD data in Extended Precision Floating Point Format, which is wider than double precision data to prevent loss of precision *during* calculations. In addition, the 287 had Control, Tag and Status registers.

All x87 op-codes are preceded by the `ESC` op-code (binary 11011) to indicate to the main processor that the next instruction is not for it to execute. In ASSEMBLY Language, the x87 instructions begin with "F," such as `FADD` for Floating Point Addition. All memory addressing modes available to the base instruction set are also available to the FPU instructions, so that data need not specifically be placed into a numeric register. However, many of the x87 operations have the ST(0) register as an implicit operand.

Because the possibility exists to use the numeric registers as a stack, it is important to distinguish which x87 instructions increment the 'stack pointer' (thereby changing the mapping of physical registers to ST(x) labels) and which do not. For example, the `FLD` instruction *pushes* its specified operand onto the register stack; the FPU's stack pointer is incremented before the push, so that the registers are remapped. To clarify this, suppose that register 6 is the current ST(0). The instruction `FLD` *memloc* loads the floating point data stored at *`memloc`* into numeric register 7 which also becomes the ST(0) after execution of the instruction (recall that ST(0) is an implicit operand, so that in effect the instruction is `FLD ST(0), memloc`). Numeric register 6 is now referenced by ST(1). On the other hand, the store instruction FST copies the value of ST(0) to the destination (register or memory) without changing the stack pointer. Other related instructions include `FTSP` (store and pop) and `FXCHG` (exchange registers).

As mentioned previously, the FPU can operate on integer or BCD data. The instructions are slightly different (for example, FILD would be used for 'integer load'), but the processor must convert all data to the Extended Floating Point Format to be stored in the numeric registers. This of course can be costly in terms of cycles if the FPU is being used for integer math. Table 2.5 shows the execution times for several x87 instructions. Again, note that the newer processors provide more execution efficiency.

If the FPU is used for a simple integer addition, considerable cycles are wasted compared to using the integer ALU. The FPU equivalent to **Listing 2.1** for a basic integer addition is given in **Listing 2.6**.

Listing 2.6 x87 ASSEMBLY listing for integer addition equivalent to x86 version in Listing 2.1

```
FILD memloc1     ; load integer data at memloc1
                 ; into ST(0)
FIADD memloc2    ; integer add, using floating
                 ; point processor, ST(0)
                 ; value at memloc2
FIST memloc2     ; stores integer data in ST(0)
                 ; to memloc2
```

In the worse case scenario, **Listing 2.6** requires 281 cycles on the 286, 85 cycles on the 486 and 16 cycles on the Pentium. This compared to the data for **Listing 2.1** in Table 2.4 using the integer ALU (15, 4 and 4 cycles for 286, 486 and Pentium respectively). Clearly, the use of the FPU for integer arithmetic is a low performance option. Without care in some circumstances, compilers may produce FPU object code where integer ALU code is suitable (see Chapter 5).

Using the FPU to perform floating point math is certainly preferable compared to using software emulation and the integer ALU, but it should be remembered that floating point arithmetic is *in general* costly in terms of cycles (Chapters 8, 9 and 12 explore ways to minimize the cost in floating point intensive programs). The FPU's in modern processors do provide a number of useful functions, and can perform useful tasks in streamlined ASSEMBLY Language procedures.

The key point with the simplistic examples given in this section is to illustrate that a compiler does not perform *magic* that the programmer cannot understand. An executable program is simply a sequence of op-codes, and each requires a certain

Table 2.5: Some x87 instruction latencies for 287, 486 and Pentium processors. Register ST0 is implied as the destination operand.

Instruction	287	486	Pentium
FIADD mem16	102-137	20-35	7/4
FILD	46-54	13-16	3/1
FIST mem16	80-90	29-34	6
FMUL mem32	110-125	16	3/1

number of cycles to execute. Since the compilers are designed for use in a broad variety of programming situations, they employ generalized algorithms to parse and 'translate' the high-level source instructions to op-codes. The resulting code is not always the best for a *particular* situation. Part of the trick to writing high performance high level code is to feed the compiler source that compiles the way *you* want it compiled.

2.5 CODE EXECUTION AND TIMING

The preceding section outlined the timing involved in the actual execution of op-code instructions within the CPU. However, the discussion assumed that the op-codes were retrieved, decoded and, in the case of newer processors, scheduled. In this section, the fuller problem of code execution is briefly presented. The motivation for this discussion is still to explore ways CPU performance can be affected by the code it is running.

The execution of instructions involves several steps, each of which are performed by highly specialized circuitry in the CPU. These steps vary from chip to chip; a common model for Pentium class processors is presented here. For the Pentium, the full execution sequence can be divided into the following steps: fetch, decode, issue/schedule, execute, retire and write back.

At the end of any given clock pulse, the IP register 'points' to the next instruction to be executed. Typically, the code is stored in physical system memory, so the IP points to a RAM address; alternatively, an instruction *may* be cached (either L1 or L2). In order to be executed, the instruction has to be *in* the CPU. It helps to think of the execution units in the CPU as being programmable gates. The op-codes specify not only the data for operations, but also the specific 'gate' to use: one gate for addition, a different one for comparing registers, etc. Therefore, the first step is to *fetch* the instruction from system memory or cache. Old CPU's required considerable time to fetch instructions, but virtually all modern CPU's utilize techniques to prevent fetching being a performance bottleneck. Some CPU's fetch 32 bytes worth of instructions into a primary instruction cache by specialized prefetch circuitry.

After the op-code is fetched, it must be decoded. That is, most PC CPU's are so-called "Complex Instruction Set Computers" (CISC) meaning the op-codes can perform relatively complex, multipart tasks. Beginning with the Pentium Pro/Pentium II class processors, the CPU itself is engineered as a Reduced Instruction Set Computer (RISC) in the core, and decoding circuitry in a sense further "assembles" the CISC op-codes to RISC micro-operations. Decoding may take 1 or more clock cycles, depending on the complexity of the native CISC instruction being decoded.

Older CPU's were purely scalar processors. Instruction execution occurred in a purely serial fashion, with the possible exception of floating point operations being sent to a coprocessor. However, since the coprocessor was a separate chip, the main CPU was still purely serial. In such a situation, multiple instruction streams *could* execute (one on the main CPU and another, purely floating point stream on the coprocessor), but often the main CPU would *wait* until the coprocessor was finished. Modern CPU's are termed "superscalar" since they consist of multiple execution units for the decoded RISC micro-operations. Execution units can be categorized as integer ALU's, Floating Point Units (FPU's), data loading, data storing, units dedicated to execution of extended instructions and CPU control units.

With the development of the multiple execution units, the CPU requires scheduling circuitry. This step in the execution cycle is termed Issuing/Scheduling. The op-code must be issued to the proper execution unit. Each execution unit may have a buffer where micro-operations are queued. Issue/Schedule may be considered a temporary storage area for the micro-operations awaiting actual execution.

The execution of the micro-operations is performed by the unit assigned in the Issue/Schedule step. Due to multiple execution units within the CPU, modern CPU's have 'out of order execution' capability. That is, the RISC micro-operations do not have to be executed in a particular order, and this is a key optimization performed by the newer hardware. However, it does introduce a 'reassembly' step such as required in some parallel processing (see Chapter 12). Output from the execution of micro-operations is stored in a temporary internal area.

The reassembly of micro-operation results into output meaningful in the context of the native CISC instruction is termed 'retiring' the instruction. Retiring is performed, again, by special circuitry that reads the results in the temporary execution unit output area. Following retiring the instruction, the final task to complete execution is to write the CISC result to memory or register as specified in the CISC op-code.

2.6 PIPELINING, SPECULATIVE EXECUTION AND BRANCH PREDICTION

Older microprocessors had the drawback that the instructions had to be fetched, decoded and completely executed before the cycle could begin for the next instruction. Clearly this was costly to performance since, for example, the fetch circuitry was idle during decode or execution. Newer processors allow an assembly line type of approached called "pipelining." The pipeline system allows each subsystem in the execution cycle to stay busy a greater percentage of time. For example, as soon as an instruction is fetched and 'passed' to the decode circuitry the fetching hardware can begin the next fetch. The actual execution step is typically the longest step, so that decoded micro-operations are placed into execution queue's by the Issuer/ Scheduler.

Pipelining adds tremendous performance gains for certain types of code. Most notable is code that executes sequentially with little or no branching. Specifically, the CPU pipelines instructions *up to* `JMP` and related instructions. For purely sequential flow, it makes sense that the longer the pipeline (meaning more instructions 'fit' in the pipeline) the better performance. However, such sequential code is rare, and many useful codes expend considerable effort in making decisions. Indeed, the author recalls early descriptions of the benefits of personal or microcomputers as specifically containing references to 'ability to make decisions.' When the program comes to a branch, the decoded and queued instructions may be lost (the pipe is flushed). In such cases, the advantages of the pipeline are lost since the CPU must then stop execution while waiting for fetch, decode and issue of the next instruction.

Processor designers initially attempted to improve the pipeline – branching dilemma by adding a function called 'speculative execution.' With this feature, the CPU 'made a guess' about which branch to follow and pipelined those instructions. Of course, an alternative was to pipeline both branches, then follow the needed pipe and discard (flush) the pipe containing the unneeded instructions. Speculative Execution was an improvement over pipe flushing, but lacked any feedback to increase the chance of a 'correct guess.'

Branch Prediction was the next evolution in improving the pipeline performance in the face of branch possibilities. With this feature, the processor actually has a small internal 'memory space' called the Branch Target Buffer (BTB) in which branch results are stored. Each time the CPU comes to a branch, the actual logical result (true or false) for that branch is stored in the Branch Target Buffer. This allows the CPU to maintain a running tally of specific branch results, so that the 'guess' as to which branch to follow is not purely speculative. Branch Prediction algorithms have also evolved with CPU architecture, so that newer CPU's perform better than older ones.

One key point to emphasize about using Pipelined CPU's is that maintaining some control of branching in high-level source code will *always* improve performance. In other words, low performance code results when unnecessary branches are used, or when algorithms are poorly designed with respect to decision structures (such as `If…Then…Else`). For example, **Listing 2.4** may be inefficient since it repeatedly follows the `JNZ` branch with greater latency (that is, It Jumps to `mulloop` more often than continuing to the next instruction). A possible improvement to **Listing 2.4** is shown in **Listing 2.7**.

Listing 2.7 **Alternate decision structure for Listing 2.4**

```
        XOR   AX, AX           ; common way to zero a
                               ; a register
        MOV   CX, memloc1      ; CX is "counter"
                               ; register
                               ; memloc1 holds 2
mulloop:

        JZ    muldone          ; Jump out if CX = 0
        ADD   AX, memloc2      ; address of this
                               ; instruction is
                               ; "mulloop"
```

```
                            ; memloc2 holds 6
     DEC   CX               ; decrement the CX
                            ; register
                            ; Z flag set if
                            ; DEC CX results in 0
     JMP   mulloop          ; unconditional jump
                            ; to mulloop
muldone:

     {next instruction in program}
```

Though this code follows the 'don't jump' branch more often than the 'jump' branch (for JZ), the JMP has been added. This increases the overall latency of the 6*2 multiplication to 56, 18 and 12 cycles for 286, 486 and Pentium, respectively; clearly, this is less efficient code than that shown in **Listing 2.4**. This is mentioned again in Chapter 4 in regard to Compiler Optimizations.

It is true that pipelining and branch prediction both improve performance, but branch mispredictions and pipe flushing do occur in a heavily branched program. Many Scientist Style programs blindly use decision structures where better algorithms should be considered. Section 10.3 shows a method for eliminating decision structures in some circumstances. Further, looping always includes a branch (continue looping or 'fall out' of the loop), so loops incur at least one branch misprediction penalty on the code. Chapter 7 focuses on ways to minimize the cost of loop branching.

Pipelining and Branch Prediction often seem to function as a trade-off in system performance. The P4 CPU utilizes a 20 stage pipeline (considered rather long by many system programmers), which explains part of its performance increase over the Pentium III. However, with such a pipeline, a rather large penalty occurs when a branch is predicted incorrectly. The CPU loses a relatively large amount of buffered instructions, so the cost 'seems' higher than a branch misprediction in a CPU with a shorter pipeline. This emphasizes why it is *very* important to understand what a particular benchmark measures when comparing hardware; code optimizations and code style may favor one hardware architecture over another.

2.7 OPTIMIZATION AT THE CPU LEVEL

The demonstration programs in this book utilize relatively large procedures (achieved by wrapping the interesting code with a loop) to require 'macroscopic' time scales. To measure the execution times, therefore, the codes simply use calls to the course Operating System timer function that generally returns system ticks in milliseconds. This was done to illustrate a very simple, though low resolution, approach to hand profiling actual working code. Hardcore profiling, however, requires direct access to higher resolution details, and these are available with Pentium and later x86 based processors.

The keys to these high-resolution metrics are the Time Stamp Counter (TSC) and the 38 internal counters accessible via Machine Specific Registers (MSR) 11h, 12h and 13h. Note that the information contained in these registers is, as the name implies, machine specific; hardware 'portability' is not an option. However, when developing for a particular architecture, the information provided can guide a programmer to dramatically improve the code. The 64-bit TSC contains the number of CPU clock ticks since the computer CPU was powered-on. Reading this register with the RDTSC ASSEMBLY instruction before and after a block of code yields the *absolute clock cycles* required to execute the code.

MSR 11h is a 'controller' register for accessing the CPU level counters. The counters available in the Pentium processor are:

Data Read	Data Write
Data TLB[a] Miss	Data Read Miss
Data Write Miss	Write Hit to M or E State Lines
Data Cache Lines Written Back	Data Cache Snoops
Data Cache Snoop Hits	Memory Access in Both Pipelines
Bank Conflicts	Misaligned Data Memory References
Code Read	Code TLB[a] Miss
Code Cache Miss	Any Segment Register Load
Branches	BTB[b] Hits
Taken Branch or BTB[b] Hit	Pipe Flushes
Instructions Executed	Instructions Executed in *v* Pipe
Bus Utilization	Pipeline Stall by Write Backup
Pipeline Stall by Data Mem. Read	Pipeline Stall by Write to E or M Line
Locked Bus Cycle	I/O Read or Write Cycle
Noncacheable Mem. References	Address Generation Interlock
Floating Point Operations	Breakpoint 0 Match
Breakpoint 1 Match	Breakpoint 2 Match
Breakpoint 3 Match	Hardware Interrupts
Data Read or Data Write	Data Read Miss or Data Write Miss

(a) Translation Look-Aside Buffer
(b) Branch Target Buffer

While many of these counters are technical and may seem of limited use to general programmers writing code in high-level languages, several are noticeably useful. For example, the Branches, BTB Hits and Branch Taken or BTB Hit counters may reveal an inefficient decision structure in the code. In addition, consider the Bus Utilization counter, which shows how efficiently the CPU is 'fed' data from memory. The Data and Code Cache Hits may reveal poorly ordered op-codes that mean the cache's are not used efficiently. Poor data organization in memory might be revealed by the Misaligned Data Memory References. Other useful counters include Pipe Flushes and Floating Point Operations. The use of these counters to profile code can lead to significant performance improvements.

The counters can report their data in an absolute hardware events mode or in a 'time spent' mode (in units of CPU cycles). That is, the Branches counter can return the number of branches or the time spent branching. This allows the programmer to fine tune the type of performance monitoring needed for a specific section of code on a specific architecture. Tuning software using the CPU counters once again illustrates that the highest performance is not possible with code that is hardware portable.

2.8 A BRIEF HISTORICAL SUMMARY OF DESKTOP COMPUTER CPU'S

This section gives a *brief* summary of the evolution of the modern desktop processors. The modern examples are easily classified as high performance units, certainly when compared to the supercomputers of only ten years ago. Much of the information regarding the Pentium (and similar) processors was adapted from http://www.tom.womack.net/x86FAQ/faq_cores.html.

2.8.1 Intel Processors

The basic instruction set and hardware architecture of modern x86 CPU's was provided in the 8088. The 8088 provided 8 bit registers, but they could be used in 16 bit pairs. All of the index and segment registers were 16 bits (providing 20 bit addresses). The chip operated only Real Mode, meaning no 'virtual addressing,' and therefore had no CPU support for a multitasking OS. The 8088's typically operated at 4.77 MHz.

The 80286 was the next major CPU in the x86 ancestry (there was an 80186, but it was not widely used in PC's). The 286 utilized 16 bit offset addresses with shifted segment registers that also allowed access to 20 bit addresses. One thing that distinguishes the 286 from the 8088 is the "Protected" operation mode, which in theory was suitable for multi-tasking operating systems requiring 'virtual memory.' 286's ran at ½ of the bus clock speed, generally 12-20 MHz. This chip never grew into its potential for three major reasons: the principle OS at the time, MS DOS, was Real Mode only with its own addressing limitations; most programs were written to be downward portable on the 8088, so the 286 features were rarely exploited; the 32 bit 80386 was developed and available before the 286 was fully utilized.

The 80386 introduced 32 bit addressing to the PC architecture. Also available were 32 bit registers. As such, the 386 was more robust for Protected Mode operation. Its popularity grew due to faster clock speeds (say, up to around 50 MHz), but most code was still written to be usable on the 8088 and 286 machines. In addition, the Real Mode OS was still most widely used. However, Protected Mode custom OS's were readily available (such as VMIX and Desqview), and MS Windows began to grow into more than just a DOS overlay.

The 8088, 286 and 386 did share one important common trait: only integer math was done on-chip. The FPU's were external "coprocessor" chips (that generally ran at ½ the main CPU clock speed) labeled 8087, 80287 and 80387 respectively. A system requiring floating point operations without a coprocessor present (and coprocessors were far from standard at that time) had to rely on libraries of software FPU emulation code. Such emulation was very slow when compared to a hardware FPU.

In the 80486, Intel put the FPU on-chip, thus standardizing the presence of an FPU. Therefore, the user did not need to consider additional hardware for machines designed for floating point intensive computations. The 486 also introduced the clock multiplier, allowing the CPU to run faster than the system (bus) clock. For example, a 486 DX2 with a 33 MHz bus clock ran at 66 MHz.

The time of 486 popularity saw the major switch from Real Mode MS Dos to Protected Mode MS Windows 3.x, though the OS still utilized 16 bit addressing (offsets) and the 486 was a 32 bit chip (this again exemplifies the lag between hardware and software advances). The 486 was probably the first PC CPU with enough 'power' to effectively utilize a multitasking OS while providing the user with acceptable response. The 486 remained popular for several years, even after the Pentium was available. At the time of this writing, 486 machines can still occasionally be seen in home systems.

The Pentium (which might logically be called 80586) marks the point where PC marketing began in earnest for the home market (prior to the Pentium and developments in MS Windows, generally only enthusiasts had computers at home; the casual user was the focus of the new marketing effort). The Pentium was purported to be "so incredible," who would want any other system? The chip is 32 bit and offered improved instruction timing, but when running 16 bit Windows 3.x did not live up to the hype. Performance increases over 486's were initially not seen as being worth the extra money by many consumers, but on the technical side, Pentium did offer more efficient pipelines and register pairing. On the negative side, the early Pentiums ran very hot, and for the first time, CPU heat was a significant issue on a "home" computer. The popularity of a 32 bit OS (Windows 98) allowed the Pentium to spread its wings, and clock multipliers were increased allowing larger effective CPU clocks.

The popularity of the MS Windows 95 – Pentium combination, along with the increasing accessibility to the Internet, led to developments in the handling of large datasets. Examples include graphics and 'multimedia' intensive Internet applications. So, CPU developers began designing towards more usability by the casual user, and the core x86 instruction set began seeing major additions in the form of MMX instructions.

The original Pentium processor included two integer pipelines. The first chips had maximum clock speeds of 66 MHz, but 200 MHz was attained by Pentium processors. Early chips ran very hot and contained the hardware `FDIV` bug, which was fixed while the Pentium was still in production. The Pentium had rudimentary branch prediction capability, though there exist some questions about its effectiveness in general code (in fact, the branch prediction capability was added as adapted from the Pentium II core while the Pentium was still being produced). The added MMX instructions provided Single Instruction Multiple Data (SIMD) parallelism for integers useful for multimedia applications.

The Pentium II (beginning with the P6 core) represents a significant improvement in the hardware design. The P6 core was designed as essentially a RISC chip with fast decoders feeding the fundamental execution units. The P6, such as marketed as the Pentium Pro, also had a three long integer pipelines, improved branch prediction and out-of-order execution (by multiple execution units). The designation Pentium II typically refers to a P6 core coupled with the MMX instructions. With this chip, the FSB clock was increased to 100 MHz and the SSE instructions were added to provide SIMD capability with floating-point operands. The Pentium II utilized an off-chip cache generally running at ½ of the core clock speed.

The significant difference with the Pentium III was the addition of 256 kB on-chip L2 cache. The Pentium III was capable of transferring 32 bytes of data every 2 cycles and boasted improved FPU performance. However, marring this latter improvement was some controversy regarding how the floating-point operations were improved; some asserted it was merely the result of highly optimized benchmarking software. Finally, the Pentium III hit a performance wall related to the manufacturing process that led to a recall of the 1133 MHz P3.

The Pentium 4 brought features to combat Intel's most significant competitor in the PC CPU market (AMD, with its Athlon line; see below). The P4 core uses a large queue, very deep buffers and a 20 stage pipeline. The chip utilizes a feature called 'traced cache' to store decoded micro-operations rather than caching native (CISC) instructions. Further, the integer ALU operates at twice the speed of the rest of the CPU, and can access system RAM via a fast bus (suggesting an extremely high memory bandwidth). Another instruction set extension was introduced, the SSE2 set, that provides double precision vector floating point operations. This set, properly utilized, may render the x87 instructions obsolete, at least for certain types of data. However, the P4, with its very long pipeline, pays a relatively large penalty for branch mispredictions, and the original bus design has proved to be less advantageous than previously thought according to some observers.

2.8.2 AMD Processors

Intel has had several competitors for chips running the basic x86 instruction set. Of these, Advanced Micro Devices (AMD) is most notable since they have offered very high performance CPU's often for less money than the Intel processors. The AMD chips have followed a similar evolution to those of Intel, but there have been technical differences that in part explain the improved performance/cost ratios. For example, it is widely thought that recent AMD FPU's consistently outperform the Intel counterparts.

AMD's answer to the Pentium (P5 core) was the K5 series. Like the Pentium II, the K5 was designed as a RISC computer with decoders to convert the CISC instructions. The K5 utilized 4 distinct execution units: two integers, one floating point and one load/store unit. The K5 was an early low cost alternative to the Pentium, but note that it included Pentium II generation technology.

The K6 was also a post-RISC via translation design. This chip employed seven execution units (load, store, 2 integers, floating point, branch and MMX), but used a non-pipelined FPU. The 3dNow! instruction extensions (a subset of SSE that allowed SIMD with single precision floating-point data) were added, and the chip was designed to operate with a 100 MHz FSB.

With the Athlon (K7) core, AMD really became noticed in the x86 market. Athlons were relatively easy to 'overclock' for enthusiasts eager to push their hardware to its limits (compared to the Intel chips that sealed multipliers so only FSB could be modified), and were generally cheaper than the Pentium III's with which they competed. There exists many web sites containing performance comparisons of both stock and overclocked Athlon systems to Pentium III's and the consensus among the high performance computing community is that Athlon beats the P3 'hands down, especially if cost is a factor.'

Technically, the chip is a second-generation post-RISC design utilizing a fully pipelined FPU. There is also a very large L1 cache, the FSB could run at 200 MHz and the L2 cache was moved 'on-chip.' The cache designs helped the Athlon beat the 1000 MHz barrier that gave the Pentium III trouble. Further, SSE instructions were added as well as hardware prefetch circuitry.

Though a competitor in the x86 market, it is interesting to note that the K7 processor shares technology with the G4 (MPC7400), licensed by AMD from Motorola. Indeed, Jon Stokes (http://www.arstechnica.com/cpu/1q00/g4vsk7/g4vsk7-1.html) made the following comment: "...regardless of marketing spin, stone-age controversies between platform factions, and general hysteria surrounding the terms "RISC" and "CISC", the K7 and the G4 are remarkably similar. They face similar problems, and they solve them in similar ways." Stokes further noted that supporting legacy x86 code (specifically, the 8 and 16 bit instructions) was one thing holding the K7 back; without this downward compatibility, it is interesting to surmise what the Athlon 'might have been.'

2.8.3 Cyrix Processors

Cyrix also provided x86 architecture chips to compete with Intel. Though there were several chips offered by this manufacturer, only two are considered here. The first is the 6x86. This chip included two execution units, no pairing rules and no FPU pipeline. It was quickly learned that this chip was not suitable for floating point intensive applications. The Cyrix3 used a pipelined FPU, included the 3DNow! instructions and operated with a FSB of 66-133 MHz.

In 1999, Cyrix was acquired by VIA. Since that time, a new chip was introduced: the C3. According to VIA itself, the chip is not intended for a floating point intensive market. In fact, according to VIA's own literature:

> *"The VIA C3 processor has not been designed or optimized for high-end graphics intensive applications that usually rely heavily on floating point calculations. VIA do not believe that performance in these applications is an important factor for the VIA C3 processor's target markets."*

There is apparently no current processor by VIA that is suitable for high demand numerical computations.

2.8.4 Motorola/IBM Processors

It is with great fondness and nostalgia that the author opens this section with mention of the 6809 processor. This chip from the early 1980's was the basis of the TRS-80 Color Computer II (CoCo), as well as a dual processor Commodore. The 6809 was an 8 bit chip with dual accumulator registers that could be paired to perform 16 bit integer arithmetic. In addition, there were 16 bit index registers, and the chip had *many* addressing modes. Indeed, some of these modes were quite abstract, and in the hands of expert programmers, very sophisticated code could be written.

The author's earliest forays into ASSEMBLY Language were with the 6809 and its chief competitor, the Zilog Z-80. For example, a low-level printer driver was written in ASSEMBLY for the 6809 to dump pixel mapped quantum chemistry orbitals to a Panasonic dot matrix printer. The 6809 was capable of performing an 8 bit integer add in three clock cycles (and clocks were 2 MHz or faster), completely killing the seven cycles required on the Z-80. The 6809 was one of the first microprocessors to contain an integer `MUL` instruction in its instruction set; this instruction was certainly absent in the Z-80's ISA.

The 6809 gave way to the 68000, one of the most successful (and still used) basic microprocessor architectures. The designator 68000 encompasses a family of processors, sometimes collectively called m68k or simply 68k. These processors, very early CISC designs, were used in Amiga, Atari, Apple Macintosh and Sun Microsystems desktop computers, as well as newer Personal Digital Assistants (PDA's). Indeed, newer models in the family are used in current Texas Instruments calculators such as the TI-89. That's a long lifetime for a processor family introduced around 1982!

The base model operated at 8 MHz, which was quite fast at the time of its introduction; however, the chip suffered from a 7-8 cycle per instruction latency.

Introduced at a time that the chief competitors were 8 or 16 bit processors, the 68000 contained 32 bit registers, but interfacing to the rest of the system was via smaller buses for the early models. The 68000 contained 8 general purpose and 8 address registers, which made it register rich compared to its Intel counterparts at the time (recall that the 8088 and 80286 only had four general purpose registers). The 68000 was multitasking capable, an ability not matched by Intel until development of the 80286.

Several generations later, the Motorola lineage developed into the MPC7400 family, which when coupled with a chipset is generally known as the G4. Note here a key difference between the systems marketed by Apple compared to the 'PC' systems: Apples are somewhat hardware homogeneous (G4 = MPC74xx + specific chipset), whereas PC's can contain a wide variety of hardware (variety of processors with a variety of chipsets available for each processor). The 7400 is a superscalar RISC processor in the PowerPC family. Early models generally operated 350-450 MHz, and included both SIMD and multiprocessor capability. The SIMD implementation in the 7400 is called AltiVec, and utilizes a 128 bit floating point vector processor.

2.8.5 64 Bit Processors

The move to 64 bit-computing has begun for workstation computing. The principle advantage of 64 computing is a much increased address space; theoretical physical memory can be as large as 32 terabytes. This is especially important when employing Look Up Tables to reduce floatingpoint processing (see Chapter 9). At the time of this writing, there are five 64 bit architectures readily available for desktop computers.

The Intel 64 bit offerings include the P4 6xx series and the 64 bit Xeon. Both are based on Intel's 64 bit ISA termed EM64T, operate with a 800 MHz half duplex system bus and support 32 bit x86 code. The server product Xeon has generally been outperformed by AMD's Operton, but a new core released in 2005 has benchmarked favorably against the Opteron. Athlon64 and Opteron are the products offered by AMD. These chips double the number of registers (reducing the low register penalty discussed briefly in Chapter 12), utilize a hypertransport full duplex 2000 MHz system bus (with the memory controller on-chip rather than provided by the main board) and run native x86 32 bit code *without performance penalties*. Both chips support and expanded SSE2 instruction set and a subset of the SSE3 set. The IBM product termed Power G5 or PowerPC 970FX drives the new generation Mac computers.

2.9 MODERN PHYSICAL MEMORY ARCHITECTURES

System memory can exert numerous influences on the overall performance. From the hardware perspective, there are currently three major memory subsystem designs: standard Synchronous Dynamic Random Access Memory (SDRAM), Double Data Rate

(DDR) SDRAM and RAMBUS. DDR and RAMBUS are the newer technologies. The comments below attempt to generalize memory hardware structure and organization, but it should be remembered that each system has unique features (this comment includes older DRAM, Static Random Access Memory (SRAM), Extended Data Out (EDO) and Fast Page Mode (FPM) DRAM systems).

The memory modules themselves are typically arranged on boards called a Single Inline Memory Module (SIMM), Dual Inline Memory Module (DIMM) or RAMBUS Dual Inline Memory Module (RDIMM) as appropriate. The channels that accept these boards on the mainboard are generally labeled, such as DIMM0, DIMM1, etc. Though RAMBUS is not a currently competitive technology, this system is included here to illustrate an interesting approach to system memory design.

2.9.1 Basic Memory Architecture

The memory subsystem consists of three major logical components: memory banks, circuitry to integrate the blocks into a single 'pool' of memory and circuitry to integrate the memory to the system. In turn, the memory interacts with the Northbridge on the mainboard, which in turn provides control circuitry between system memory and to the CPU. The banks contain the actual memory cells, and the mechanism for reading and writing to these cells gives each of the memory subsystems its character. The technical details presented in this section were summarized from Jon Stokes' pages at ARS Technica, beginning at www.arstechnica.com/paedia/r/ram_guide/ram_guide.part3-1.html.

At this point, it is worth noting that memory reads (data moves from memory to CPU) are not equivalent to memory writes (data moving from CPU to memory). Reading memory is considered, from the hardware design perspective, a time-critical task; the CPU must 'wait' for the data. Therefore, read latency is usually given as a specification. Write latency can be higher, without loss of system performance, except in the case of back-to-back write-read operations to the same address. It should be noted also that these memory latencies are not related to CPU instruction latencies. A CPU `MOV EAX, memloc1` instruction may theoretically execute in one cycle on the CPU, but the memory read latency might certainly exceed that.

The fundamental unit in memory is a cell, which stores one bit of information (a 1 or 0). To the CPU, and indeed the rest of the system, individual bits are not typically accessed in memory; rather, bytes (8 bits) are considered the fundamental logical unit. The physical memory system, therefore, must organize the cells into a structure that logically appears as bytes to the rest of the system. This organization occurs in the *modules*. A RAM module contains cells organized in a grid (row and column) structure, with the cells grouped into bytes. A specific *byte address* is specified to the module by setting the CAS (Column Address Select) and RAS (Row Address Select) signals.

The modules generally interface via 8 bits. In Pentium systems, the data bus is 64 bits wide, so that eight modules connect to the bus. The addresses within these eight modules are structured such that Module0 holds Byte 0, Module1 holds Byte 1, etc; a given module stores every eighth byte. This is important to improve system performance since within any given module, only one row can be active at a time, and there can be considerable latency in canceling the selected row and selecting another. If a module contained successive bytes, and requested addresses cross a row boundary, the read would have a large latency due to the change in RAS. With the 'round robin' structure, *each module* has an active row, so successive addresses are less likely to cause a RAS change. The group of modules is a Bank.

The module/bank structure also has a significant consequence in how memory should be added to a system. For SDRAM, the eight modules are physically located on one bank, and a given DIMM may contain, say four banks. To illustrate the consequences of this, consider a desired physical memory space of 1 gigabyte. A single 1 GB DIMM with four banks can have at most 32 *active* rows (one row in each module with eight modules per block). In contrast, using two 512 MB DIMMs to put 1 GB in the system gives 64 *active* rows, since each DIMM has four banks. Two DIMMs doubles the number of active rows, which decreases the RAS latency penalties for row address switches. This may result in notable system performance differences for large, memory intensive applications.

The details of SDRAM hardware are quite different from the older memory system designs. Specifically, there are two key differences of note: memory clock signals and the application of wait states. These differences set the stage for a brief outline of the SDRAM, DDR SDRAM and RAMBUS technologies.

2.9.2 SDRAM and DDR SDRAM Memory

Older memory systems, such as DRAM, did not use the system bus clock to provide pulses for commands. In some cases, synchronization was required between the system bus and the execution of the memory commands at the hardware level. With SDRAM, designers constructed the system to share the system bus clock with the CPU and chipset. This significant design feature allows memory subsystem commands to be placed on the leading edge of a clock pulse.

The older systems also had latency problems with fast CPU's that required the CPU to insert 'wait states.' In particular, if there was a CPU – memory clock speed mismatch, the considerable latency of the memory required the CPU to *wait* for data from physical RAM; during the wait cycles, no work was done. The performance penalty was largest for faster CPU's, since the speed mismatch was greater. This in part explains why early advances in processor clock speeds led to disappointing performance with real applications; true, the CPU clock speeds were faster, but the CPU sat idle while waiting for data. The solution with the SDRAM design was to place the wait states, when needed, in the RAM hardware, rather than the CPU. With this improvement, the CPU need not sit idle while waiting for data (though it may, in some circumstances).

DDR SDRAM was an evolutionary improvement of the basic SDRAM system by placing memory subsystem commands on both leading and trailing edges of the bus clock signal. As a result, data access rates were doubled. There is a little extra hardware circuitry required on DDR modules, but the added overhead is small for the performance improvement provided. At the time of this writing, DDR bus clocks exceed 400 MHz which 266 MHz and 333 MHz buses remain common.

2.9.3 RAMBUS Memory

Like many technological improvements in PC computer design, the RAMBUS architecture was heavily marketed as a revolutionary step that would virtually render other technologies obsolete. RAMBUS is not an evolution of the SDRAM design, but rather a complete redesign of the memory system. However, again like many design advances, actual experience differs from the hype. The RAMBUS does contain many design advances, and certainly provides a lot of potential for further improvements, but currently suffers from three major drawbacks: power consumption, complexity (which effects cost) and the ramifications of having a very long, serial data channel. It should be noted that these drawbacks are not always manifested in a given system, so that RAMBUS *may* significantly outperform DDR SDRAM.

RAMBUS is fundamentally different from SDRAM. Each RDRAM chip holds 32 banks, half of which can be 'active' at any given time. Unlike the banks in SDRAM, the RDRAM banks are not spread across modules but are complete units in themselves. This means that for a given physical memory size, the number of banks (and thus the number of possible active rows) is very large. RAMBUS achieves very high theoretical memory bandwidth by having a large number of banks. The drawback of the large number of banks is very high power consumption.

RAMBUS also has an increase in overall hardware complexity. For example, the individual RDRAM chips must be multiplexed/demultiplexed to the RIMM interface. Multiplexing is a means to convert parallel data to a serial stream; in this case, 8 bit parallel data is converted to 8 serial bits in a read operation; demultiplexing, the conversion of 8 serial bits to 8 bit parallel data, must occur during write operations. The overall row/column address decoding circuitry is more complex for RDRAM than for SDRAM. Further, not only are the individual memory chip structures more complex, but the chipset is as well. Indeed, the chipset must also perform multiplexing/demultiplexing operations before data is sent to or from the CPU.

As described, the fundamental hardware is very different for RDRAM and SDRAM. SDRAM is a system using a parallel data channel at relatively low clock frequencies. In contrast, RDRAM uses a serial data channel (between chipset and memory) running at very high clock frequencies. In other words, SDRAM has the connectivity complexity engineered into the bus lines between memory and chipset, where RAMBUS places this complexity not in the bus, but in the components themselves (RDRAM chips and chipset).

The added hardware complexity has a significant performance drawback that must be noted. RAMBUS does not use the system bus clock for internal operations. Therefore, the RDRAM pulses must be synchronized to the system bus. Such synchronization can add latency to the overall system. This is mentioned since much of the marketing and popular media extolling the virtues of RAMBUS deal mostly with theoretical performance of the memory subsystem itself. The theoretical bandwidth is often not realized due to integration issues with the overall system.

Data in RDIMM memory systems follow a single, serial channel. The single channel has several consequences on performance. Each signal from the chipset must follow the entire channel, whether going to a RDRAM close to the chipset or one 'far away.' This results in latency issues, since RAM that is 'far away' from the chipset will have a high absolute latency. Complicating this latency issue, different RAM in the system has different latency depending upon location in the channel. This can have significant performance consequences. For example, suppose the CPU requests two pieces of data, *a* and *b*, in order. If *a* is farther away along the channel, the latency for transfer from *a* will apply to *b* as well; *b* cannot be transferred until the transfer of *a* is complete. That is, *b* by itself would have a smaller latency than *a*, but when *b* follows *a*, the transfer of *b* sees *a* as a bottleneck. In SDRAM, including DDR, this is not an issue since all RAM has the same latency and the bus structure is parallel.

There are several notable advantages in the RAMBUS design. For example, RAMBUS uses current mode switching to change states along the bus in contrast to voltage mode switching generally used in other systems.

In the final analysis, RAMBUS was generally found too complex and expensive. It is instructive to emphasize that several years ago, RAMBUS was to revolutionize the PC market. That Intel began making P4 boards suitable for DDR bus architecture is testament to the fact that DDR was a strongly competing technology, especially as DDR frequencies exceeded 266 MHz.

Chapter 3: Operating System Considerations

The operating system (OS) of a computer exerts considerable influence on the performance of programs. An executable is a sequence op-codes for a given CPU, so it may seem that the above statement is a contradiction if one hardware architecture is considered; for example, the op-code for floating point multiplication is determined by hardware not OS. However, most programs, especially those written with high-level languages such as C, Fortran, BASIC, etc, interact extensively with the OS unless the programmer takes great care in writing source that is OS independent.

The above statements refer to platform or OS portability. The source code may be written for a *specific* hardware (which is therefore not hardware portable), or the source code may be hardware general but not OS portable. Visual C++ is a good example of a compiler that requires source code only for MS Windows; the compiler utilizes considerable meshing with the Windows OS, and Visual C++ source code will not compile on a different OS's C++ compiler using the same hardware. Demo program *Hardware01* (from Chapter 2) demonstrates conditional compilation by including MS Windows and Linux specific code in the same source file.

However, even if portability is not an issue to the programmer, code interactions with the OS demand serious attention for high performance code generation. How the OS controls physical and virtual memory, disk access, task swapping and other system specific details will influence the execution of a program. Further, the OS provides a multitude of 'built-in' functions that may be useful for a programmer (ie, why write a function when the OS already contains it). These callable system level functions are generally termed an Application Programming Interface, or API.

In this Chapter, the general interactions between a program and the OS, OS architectures and some of the features of several common PC OS's are considered. This is not intended to be a thorough and 'complete' guide to specific OS's, but rather an introduction to some of the ways a code may interact with an OS. MS Windows and Linux are considered in some detail.

3.1 THE OPERATION SYSTEM IN PERSPECTIVE

In Chapter 2, the execution of code at the processor level was examined, but the program in an overall system context was not presented. To develop the idea of a program performing useful work, however, the entire system, not just the processor must be considered. A program must interact with memory, i/o devices such as disk drives, printers, etc and even must compete for these resources with other programs. The control of all of these interactions is performed by the Operating System.

Figure 3.1 depicts a functional chart for a basic computer system in terms of the executing "programs." In principle, the OS is not necessary since once running, an application may make calls to the BIOS (Basic Input Output System) or hardware directly. However, the OS does control availability of resources and program execution generally begins with calls to and from the OS, so the OS kernel plays the role of 'traffic cop.' To be more specific, when a program begins execution, it requests memory from the OS to load the code from disk (and the OS reduces that memory from its available pool) and the OS schedules the application for CPU resources. The OS also designates memory for data and stack space. On single process systems (such as MS-DOS), there are no shared resources; the OS essentially serves as a user interface and to provide enhanced functions.

At the lowest level, the hardware, code is present that allows the hardware to function: to turn on, to accept input, provide output, etc. As hardware is generally faster than software, this software layer is typically as 'thin' as possible to allow the hardware to 'exist' in the system. Note that this layer of software is not what is commonly called a 'driver,' and most likely exists in ROM or flash memory. Though some graphics cards formerly required direct programming of hardware (Hercules, for

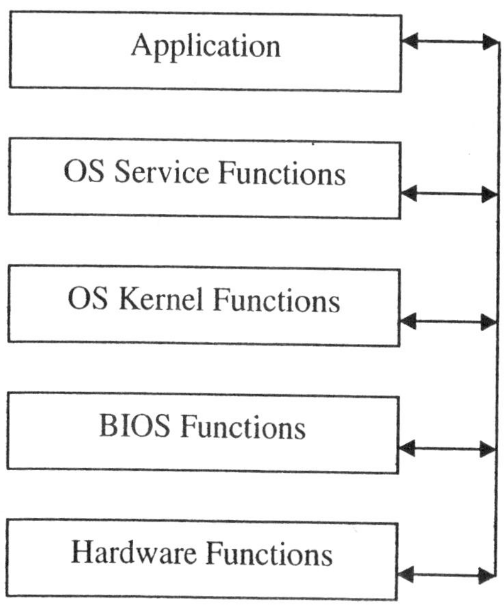

Figure 3.1: Schematic representation of the system software hierarchy running on a basic computer.

example), direct programming of hardware is rarely part of programming modern applications.

The BIOS forms the next tier, and it is here that useful functions begin for the application programmer. However, these are still low-level calls that generally require considerable programming expertise to utilize. On the PC, for example, the BIOS routines are accessible via software interrupts, generally after loading parameters into specific CPU registers. BIOS routines include keyboard i/o, screen i/o, disk i/o, etc. There is little need to access BIOS routines for general applications, but programmers of utility software often make use of them to by-pass safety codes and error checking used at the OS kernel and higher levels. For example, this author once wrote a data recovery routine, in ASSEMBLY Language, to retrieve data from a floppy disk that was producing a "General Disk Error" in MS-DOS; the program used BIOS level disk calls to create an image of the bad floppy on the hard disk. This image was then readable by other utility software. Only a relatively small quantity of the data was actually lost rather than the entire disk's worth.

The OS kernel represents a large step in the functionality available to the application programmer. For example, at the kernel level, disk i/o routines include file system structure, error checking and perhaps security measures such as file permissions. Modern OS kernels include a plethora of utility functions to facilitate system administration and software troubleshooting. Graphical OS's have object-oriented kernels so that modules and data structures can be added at higher levels without rewriting kernel routines. The kernel is that part of the OS that allows the computer to function as a system, particularly in shared environments (multi-user, multi-process, etc). Included here are software drivers for the hardware that allow the application level to communicate with the hardware via common or standardized 'interfaces.'

The OS also includes a set of service functions. These routines are not required by the system, but bring to the system efficiency. As an example, consider a print spooler. The system does not require a print spooler to operate; files can be sent to the printer singly. However, having a spooler allows one user to send a file to a 'printer' (more correctly, to the spooler) without knowing if the printer is available. This allows users to seamlessly use the system; to the user, the printer is always available (if the spooler is running!). Other service functions may include mailers, file back-up utilities, resource monitors, etc. These services may be 'applications' or they may make their functions available to other programs.

At the top of the software pile are the applications. For this purpose, the term application can be defined as a program that executes to perform useful work that is not making its functions available to other programs (the reason for this latter distinction is simply semantic; system-wide callable functions are being categorized as part of the OS for purposes of classification). Applications may contain code that is independent of the OS (once the application is running, since at the very least, an application must request execution resources from the OS) or may contain function calls to the many available OS service and kernel functions. The choice of whether to write a function or use an existing one is made by the programmer and many factors should be weighed in the decision (See Chapter 6). From the point of view of generating fast code, the question is simply "Is using the OS function fast enough?"

3.2 OPERATING SYSTEMS AND PERFORMANCE

The extreme levels of readily available built-in functions have conflicting roles. At the kernel level, functions are typically very general. This generality may kill speed. For example, the Windows Graphical Device Interface (GDI) is an API for interacting with the Windows Graphical User Interface (GUI). The Windows GDI is useful for placing information on the computer monitor screen, and code written for it will run on any computer running Windows. However, the GDI is found to be very slow, since it contains only 'broad stroke' basic functions and is not performance oriented. Programmers of 3-D graphical applications (games or scientific visualization suites) would not use the GDI for general rendering of 3-D images; these functions are simply too slow.

A simple program can illustrate the performance limitation of using the GDI function calls in Windows. Demo *OS01* uses three methods to draw and paint (fill) 10,000 rectangles in a window, and the 1200 MHz Athlon results are shown in Figure 3.2. The first method uses code to draw the outline of a rectangle then uses the Win32

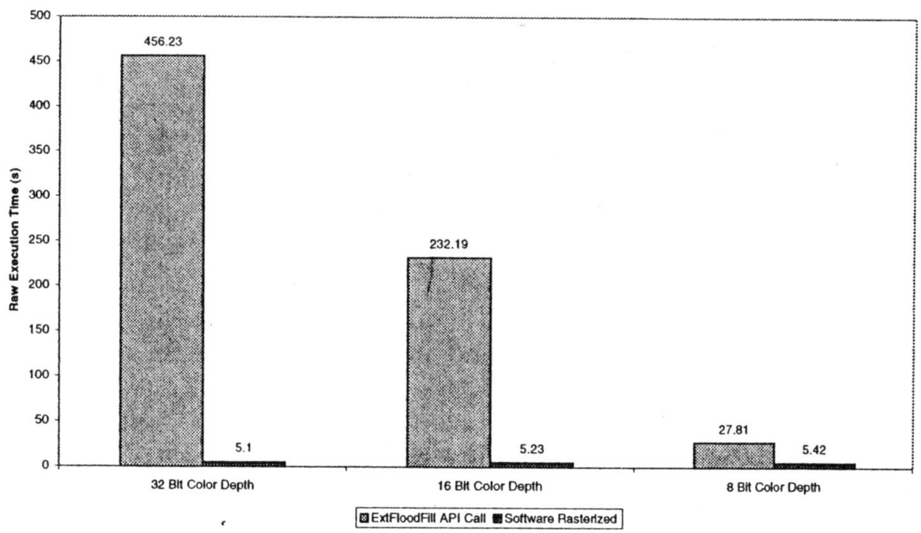

Figure 3.2: Demo *OS03* raw execution times on the test computer for drawing 10,000 random, filled rectangles on the screen in Visual Basic. Use of the API ExtFloodFill is compared to an inline, hand-coded rasterizer. Not shown in this figure, due to the small execution time, is the result for using the API Rectangle function.

API ExtFloodFill function call to fill it. On a Windows 2000 (or Windows XP) system, this call is very slow for higher color depths (note that this code was not compiled with the Windows 2000 'patch' for Visual Studio 6.0). Actually, it is still quite slow even for the 256-color (which is an 8 bit color depth) setting. The second method, which "paints" the rectangles line by line in code(using the API Line() function), is clearly much faster (this is a simplified 'software rasterizer'). Further, the hand-coded version is not dependent on color depth, so it performs equally well on higher color depth settings.

Painting 10,000 rectangles in 5 seconds (hand-coded method in Example OS01) is still extremely slow when compared to modern 3-D graphics applications that can draw millions of triangles per second. Example *OS01* draws 'flat' rectangles with ordered coordinates, no interpolation is needed, uses no clipping logic and is not painting textures or lighting parameters, so it is an overly simplistic demonstration. In addition, this example does not take advantage of any specific optimizations in the hardware (video adaptor). However, it does show the potential cost of API calling vs. hand coding. The API does have utility since many functions would only be needed occasionally and these provide useful functions to the programmer. In addition, initial testing of program logic can utilize the API calls while postponing writing (and debugging) hand-coded routines.

The API contains a Rectangle function, which executes *much* faster than either filling a rectangle with ExtFloodFill or software rasterized filling of the rectangle. This illustrates that is very important to match the needed function with the proper API call; clearly, one would not wish to use ExtFloodFill for rectangle filling. However, ExtFloodFill is a general routine that can be applied to any shape. This again illustrates how functional generality often imposes a performance cost, and this idea will be seen again in later Chapters. The call to the Rectangle API function is also demonstrated in *OS01*, which ran in 0.18 seconds on the test computer (this is not shown in Figure 3.2 since it is such a small number).

At the other extreme, hardware is very fast, but interactions need be very specific. For example, one can program the video memory directly (as opposed to using the Windows GDI for graphics) for raw speed, but each video card would require a different program (or at least a different rendering function within the application, in which case some provision for which function to call during execution must be made). Programming in this fashion also has the limitation that hardware advances will generally require a code rewrite.

It is the job of the OS to provide a usable compromise. Continuing with the Windows graphics example, this compromise comes in the form of DirectX. DirectX is a Component Object Model (COM) package that provides a software interface to the hardware allowing the programmer to essentially write to video memory. In this implementation, the OS service functions in DirectX provide the 'translations' to whatever particular video card is present. Early attempts to provide these translation functions were very slow, but as DirectX has matured (and hardware manufacturers increased compliance to standardized i/o practices), 'real time' complicated 3-D rendering became commonplace. OpenGL is another API for graphics rendering, but does not include all of the 'features' of DirectX; however, OpenGL is available for more platforms than is DirectX.

While most of this book's focus is number crunching rather than interactions with hardware, a key exception includes transferring data to/from disk. In this context, even a streamlined program must interact with the OS, especially when temporary storage is required (for example, swap files to accommodate data structures larger than available physical memory). The user of high performance code needs to understand how the program is interacting with the system (ie, via kernel or service calls, etc) to get the most efficient execution.

Consider for example MS Windows 2000 through Service Pack 1. This OS did not include functionality to recognize Advanced Technology Attachment 100 (ATA 100) hard drives, so the higher data throughput of these devices was not utilized (that is, theoretical disk i/o bandwidth was limited to 66 Mb/s rather than 100 Mb/s). This author encountered this limitation during the execution of *ab initio* quantum chemistry calculations during which CPU utilization was only about 40-60% of the total execution time. Upgrading the OS to Service Pack 2 (which contained, among other things, the patch for proper ATA 100 drive communication) improved the CPU utilization to 90-98%. Clearly this is a much more efficient use of the computer, since drive access is slower than CPU execution. Since the CPU's portion of the calculation did not change, net wall clock time decreased with the reduction in drive access time; in summary, performance was doubled.

The example just given was not intended to point out a limitation with MS Windows 2000, but rather to illustrate the importance of (1) performance profiling and (2) understanding the OS, hardware and software interactions. These concepts are equally important to scientists simply executing available codes as well as those programming their own solvers and applications. Further, the lag between software and hardware development was encountered: Windows 2000 was initially written only ATA 66 interface was common, and the OS did not 'allow' for improved IDE interfaces. It is quite probable that a 'standard' installation of whatever OS is being employed does not configure the system for optimal computational performance.

Another example of OS interaction deals with the swap file. For calculations requiring a larger memory space than the available physical memory (keeping in mind that in shared systems, the available physical memory can be relatively small), the OS writes images of the physical memory to disk. Since disk i/o is slower than physical memory data transfer rates, swapping slows overall execution. Clearly, the ideal situation is to have a large enough physical memory space for the calculation, but this is not always possible. Matrix inversion, which is used in a broad variety of scientific applications, is a common example of a memory intensive calculation.

Should a program always use the largest available virtual memory? Not necessarily since the speed of swapping depends on throughput to the disk. Letting a particular program use too large a value may be slower than specifying less memory. There is no magic swap file size or per process requested physical memory space for all systems. The Windows 2000 Resource Kit includes extensive procedures for profiling system performance, and indeed, one of the most difficult system aspects to optimize (for heavily loaded, dynamic systems) is the swap file size. Though with current high capacity disk drives, it may be tempting to make the swap file as large as possible, doing so may make the disk i/o of the swap file itself the execution bottleneck.

It should be noted before moving ahead that DirectX and similar developments in OS services have allowed computers to be accessible to a relatively new class of user. Merely ten or twelve years ago, one needed some skill to interact with the OS to operate a computer for even simple, mundane tasks. Development of OS services has added to the Plug-n-Play idea for hardware recognition, graphical interface "wizards" that perform basic tasks for the user and a myriad of other usability enhancements. Such enhancements render much of the OS technical detail invisible to the user. While many scientists operate the computer at this level, true high performance cannot be realized until execution is properly profiled and the system configured for performance.

3.3 OPERATING SYSTEM ARCHITECTURES

The OS performs many tasks in the system, and how these tasks are implemented determines much of the performance of the computer. For example, and building on the ideas of the previous section, the OS provides: kernel, services, security and defines a unit of storage (the file system). For any application running on a system, there is a certain OS overhead, even if the executable does not contain explicit OS function calls. This overhead includes, for example, security, messaging (between objects of the executable as well as other system objects), synchronization, etc. An OS also includes basic execution units and task swapping models.

Execution units include jobs, processes and threads. A job may be considered as an umbrella process that creates and calls 'child' processes as needed. Technically, a process is an execution unit that owns its own address space; each child process in a job has its own, distinct address space. Processes can create threads, which are execution units that all exist and operate within the address space of the calling process. Since system resources, specifically memory, are controlled by the OS, a job that spawns multiple child processes by definition must communicate extensively with the OS. Thread creation also requires interactions with the OS, but additional memory allocation is not required since a thread exists in the process space that has already been allocated.

Clearly with these differences in execution units, an operating system that utilizes processes as the fundamental unit will have different performance characteristics than one that fundamentally works on the thread level. The process based model derives from older main frame computers whereas threads are more of a microcomputer idea.

The OS may also be event driven, meaning it is constantly responding to signals, messages or events generated by software or hardware. Such messaging is inherent in any shared system. MS-DOS required very little messaging, since a program was started, executed and ended before another could begin ('terminate but stay resident' utilities notwithstanding), whereas Windows includes extensive messaging between objects and the kernel.

Another very important parameter for operating system performance is how it executes context switching. Technically, a context switch is performed at the hardware level in the CPU, but the OS must implement the switch. Different implementations of context switching will effect overall system performance.

3.4 SPECIFIC PC OPERATING SYSTEMS

To briefly illustrate how the performance issues outlined in Section 3.3 are implemented in real PC operating systems, MS Windows and Linux will be summarized from the system perspective. This outline is not meant as a specific endorsement of one versus the other, but merely to present to the reader some ways that the OS architecture influences computational performance.

3.4.1 MS Windows

MS Windows is a thread based OS. If one goes to Task Manager (in Windows 2000, for example), one can view a list of current processes on the system. It should be noted that rarely does a process name appear more than once. As a process requires a new execution unit, threads are created, but it is possible for processes to spawn other processes. The idea of 'job' has not typically been used in a Windows environment.

Windows is not a true multi-user system in the classical sense, as it is unusual (except on dedicated server computers running special software) for multiple users to be running programs on a single system. Further, connecting to a Windows computer and running software that does not exist on the local computer 'terminal' is likewise unusual. Another way of stating this is that in networked systems, Windows is natively constructed as a Peer-to-Peer system rather than Client-Server.

At the risk of belaboring the point, consider a network of three computers, each running Windows. To use MS-Word on any system, each must have MS-Word loaded. A computer in this network acting as a 'server' is generally a file server, providing only a repository for data files. The binary image executes on the local 'terminal' computer in its address space; all video i/o is local, between the executing image and the local video adaptor. Users at each 'client' computer (three in this example, since there is no reason the 'server' cannot behave as a client as well) run separate processes in separate address space on separate computers.

The issue is not necessarily one of native capability, but rather of licensing. To use MS Windows as a terminal server, that is, a computer that runs software and serves i/o to 'dumb terminals,' requires additional, expensive software. (http servers running scripts on the server machine are an interesting exception). In this case, the executables run on the server computer and only keyboard inputs and video outputs are exchanged between server and terminal. In this case, for example, MS-Word would only have to exist on the server computer. Microsoft currently licenses Terminal Services by how many client terminals are connected to the server, just as individual copies of Word would be licensed in the Peer-to-Peer network outlined in the preceding paragraph.

Windows is a preempting, multiprocessing OS. Numerous processes can 'run' at once, and foreground/background execution can occur. The concept of foreground in Windows is called 'focus.' Different processes can have different priorities, and so may get different proportions of the available CPU resources. In addition, Windows is message based.

In consideration of performance, Windows suffers a major drawback. The developers have tended to focus more on including basic usability features rather than making such features configurable. With Windows, one must typically take what you get. Having said this, there are well developed programming tools to write programs for Windows, such as the C/C++ compilers, that allow essentially 'by-passing' most of the Windows excess baggage. As an example of the 'inherent complexity' that is almost forced onto the programmer, consider that compiling C/C++ programs LEAN_AND_MEAN (without Microsoft Foundation Classes, MFC) generates high performance, fast code, but this requires some expertise if the program is to have a 'common' Windows graphical user interface. The expertise is required since the programmer must create and register the Windows Class at least for the application itself. One would think the opposite to be true: simpler, less encumbered code *should be* easier to generate!

One *can* generate 'console' applications in Windows that do not contain the Windows overhead at all (the creation of the console window is done automatically by the OS). Demo's *OS02*, *OS03* and *OS04* are three programs that perform the same computational task using three Windows C/C++ programming paradigms: MFC application, simple Win32 application (no MFC) and Win32 console application. Each was compiled using the compiler optimizations, and raw execution data are shown in Figure 3.3(a). For a program involving a sophisticated user interface, MFC would be the 'easiest' and quickest of the three approaches, while the console application is the simplest program. However, the data in Figure 3.3(a) clearly show MFC to carry significant, performance lowering overhead, and the console application is slightly faster than a straight Win32 application. Therefore, performance programs for Windows should probably be written as console Win32 applications, especially for 'background' type numerical programs that do not require a sophisticated user interface. When a user interface is needed, one can generate user interfaces in non-MFC C/C++ Win32 applications, but it does require considerably more effort than when using MFC. (Since the demo's throughout this book are differential comparisons, that is comparing two different *functions* within a program, they were written with

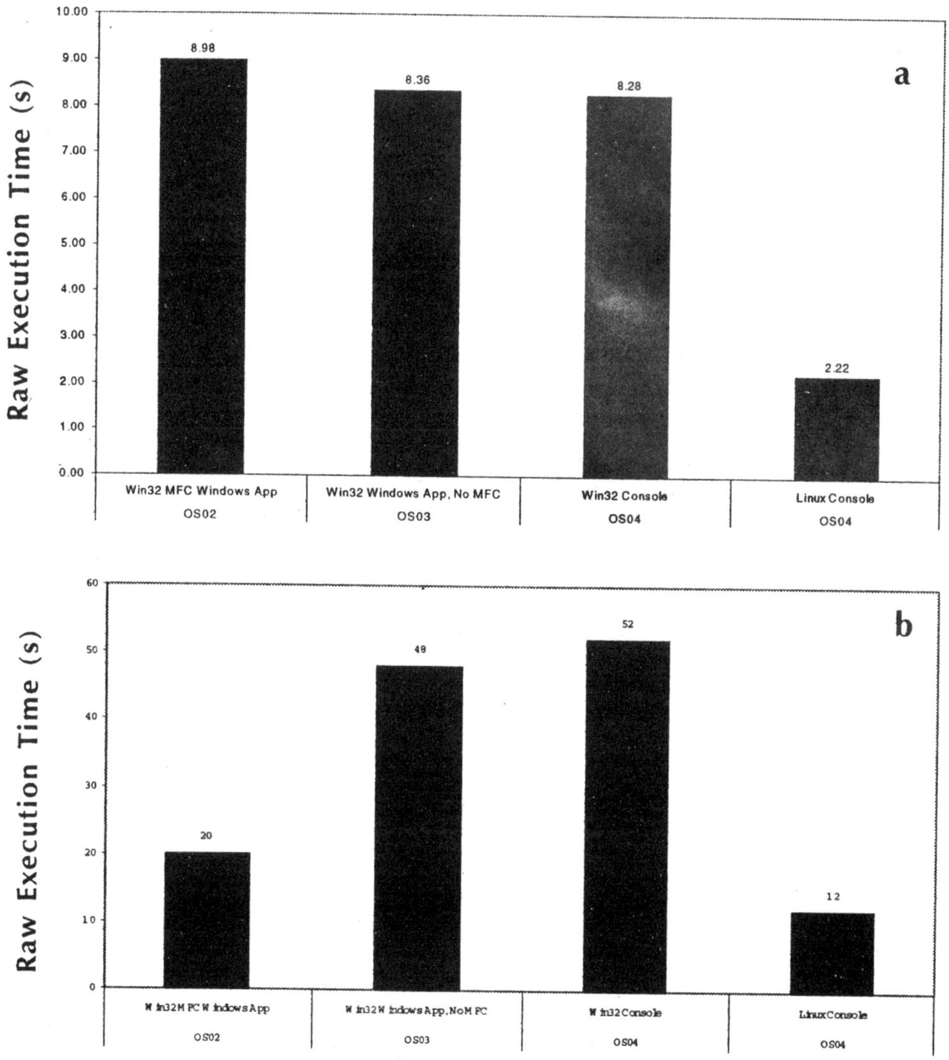

Figure 3.3: Raw execution times on the *same* test computer and compiled executable file size for different 'versions' of the same program. The code selects a random distance and computes the Lennard Jones pair energy for two Helium particles. Three Windows compilations are compared: MS MFC application, a simple (non-MFC) Windows application and an MS console application. Also shown is data for a Linux console application, again on the same computer. Note that both compiler optimizations and source code optimizations from Chapters 6 and 8 were used in this demo.

MFC to allow the quick generation of the user interface; the author would not use MFC for a high performance application).

In Figure 3.3(b), the compiled executable file sizes are shown. It is interesting to note that the MFC application is the smallest of the three Windows versions. This reinforces the suggestion of tight integration with OS for the MFC application; much of the overhead the MFC application 'uses' is external to the application itself. It is interesting that the console application is the largest. Windows executables are stored in the 'portable executable' (PE) format. Some streamlining of the program in this format is possible.

The execution times for Demo's *OS02*, *OS03* and *OS04* serve to show what seems to be a general principle with MS Windows. Difficult tasks are relatively simple and simple tasks are relatively difficult. This makes sense when one considers for whom Windows is generally intended: a so-called casual computer user. MS Windows has evolved into an OS that users with virtually no experience or training, nor the desire to obtain much of either, can use a computer *productively*. To this goal, it's probably fair to say that Microsoft has succeeded tremendously; there are people effectively using computers that ten years ago would have likely denied trying. The goal of Windows must be kept in mind when programming for performance is considered. Windows *can* be programmed for high performance, but it takes more skill (and research) than simply writing generic, portable source code.

Windows is often called a 'closed source' system. In this model, the end user purchases executable binaries or rather the license to use them. All system configuration capability is determined at compile time (by Microsoft programmers) and thus cannot be changed. In addition, system programming for the closed source OS involves either much "reinvention of the wheel" or reliance on function calls to existing modules. There is no middle ground. Recent history shows this to be an acceptable situation for many casual computer users, but high performance users require additional flexibility. Without the user's capability to experiment, OS advances are made only at Microsoft, and often finding documentation to bugs and errors is difficult. Further, Microsoft charges a fee for technical support regarding the OS, and this seems to include reporting bugs in the system.

MS Windows stores OS configuration information in a hierarchical database called "The Registry." Either loved or hated by system users, the Registry is often thought to be a cryptic approach to configuration information the user should have as readily available. Indeed, earlier versions of Windows (pre-Windows 95) used editable configuration files typically given the .ini extension; the main one of these was system.ini. Conversion to the Registry system is thought to provide a centralized location for configuration data so that programs can update the Registry as needed upon installation. The Registry data are known as "keys."

Often, the default settings for many of the keys are not optimal for performance, and system tuning inevitably involves editing the Registry. A simple example is the "Last Accessed" date/time stamp written for files in Windows 2000. In a default Windows 2000 installation, this date/time stamp is active, so that *every time* a file on disk is accessed, the OS writes the current date/time in this field of the file's entry in the directory. This takes time since system calls must be made to *get* the current time as well as *writing* the time data to disk. Large computational programs that write output files as they execute are continually interrupted so the OS can update this information. Beyond some arcane security checks, knowing the last date/time a file was accessed is of little value on a computation oriented system (as opposed maybe to servers and other network applications), and a performance gain in the system overall will be realized by turning it off.

While a number of Windows configuration settings can be changed via dialog boxes in "CONTROL PANEL," and other similar ways, the Registry is typically edited with a utility program such as regedit.exe. For example, using regedit, the key to disable the Last Access date/time stamp in Windows 2000 can be changed from the default. The key itself is (or is similar to)

HKEY_LOCAL_MACHINES\SYSTEM\ControlSet001\Control\FileSystem\NtfsDisableLastAccessUpdate

and has a default value of 0 (meaning the LastAccessUpdate is *NOT* disabled-please note the double negative). Changing the value of this subkey to 1 will prevent the OS from constantly updating Lass Accessed date/time stamps.

Loading programs at system start-up (or user log-in) is another context in which the Registry settings may hamper performance. Loading of unneeded components uses memory, and depending on the quality of the program, may cause CPU overhead or memory leaks that further degrade system performance. Many programs offer the user an option to disable automatic start-up loading; however, some do not. For example, the Hewlett Packard 990 series DeskJet print utilities automatically load, and there is no way to prevent this other than editing the Registry. The Registry contains several keys that determine program loading at start-up, such as those listed in HKEY_LOCAL_MACHINE\Software\Microsoft\Windows\CurrentVersion\Run (and there are other, similar locations in the registry). To disable automatic loading, simply remove the entry for the unwanted component.

Table 3.1 contains a partial list of Registry settings that may be useful for system performance tuning. This list is not intended to be all-inclusive; additional information regarding custom Registry settings can be found by searching the Internet, but it is recommended that caution be used in editing registry settings for which the user does not have specific documentation. Any time the Registry is changed, the old values should be saved and any system changes should be documented in a System Administration log. Of course, the system back-ups should be current before editing the registry. Further, improper registry parameters may crash Windows completely so that the only recovery is a full operating system install.

Windows does make available to the programmer many of functions that comprise the OS. While WinAPI programming is beyond the scope of this book, a small list of some useful API functions is shown in Table 3.2. The choice of whether to use these functions or write one's own is best made within the specific programming context.

Table 3.1: A list of some useful Windows 2000 Registry Keys. *Change Windows Registry Settings at your own risk, and only after appropriately backing up the system. Some registry failures are unrecoverable and require reinstallation of Windows!* The values to assign the keys are generally not given in the table, since sometimes they will be system dependent.

Registry Key		Description
HKEY_LOCAL_MACHINE\ SYSTEM\ControlSet001\ Control\(or similar for other control sets)		
	SessionManager\MemoryManagement \ClearPageFileAtShutDown	This security setting causes the pagefile to be overwritten on system shutdown, thus preventing an intruder from reading data via the pagefile. Not a performance tweak.
	Session Manager\MemoryManagement \DisablePagingExecutive	Prevents the OS from paging executive files to disk, forcing them to stay in RAM. For systems with adequate physical memory, significant performance increase is possible.
	Session Manager\MemoryManagement \LargeSystemCache	System puts as much of OS kernel in memory as possible; this improves kernel performance at the cost of memory usage.
	FileSystem\ NftsDisableLastAccessUpdate	Set to 1, disables the "Last Accessed" file date and timestamp
HKEY_LOCAL_MACHINE\ SYSTEM\ControlSet001\		
	Control\Session Manager\Memory Management\LargeSystemCache	Specifies whether the system favors the system cache working set or processes working set
	Services\Tcpip\Parameters\ Interfaces\[InterfaceID][a]\MTU	Sets the Maximum Transmission Unit for a tcp/ip network, which can improve network performance.

a. [InterfaceID] is a label for a specific Network Interface.

3.4.2 Linux

Linux, a very widely available Unix-like OS for PC's, began in the 1990's. The command structure, file system structure and execution models of Linux bear more than a passing resemblance to Unix. Like Unix, therefore, Linux is a process based OS. Jobs spawn additional processes as needed, so the smaller execution units have their own address spaces that must be maintained by the OS. Since it is common for a job to spawn additional processes, the process list (via the "ps ax" command, for example) may contain multiple processes with the same name (though each will have a unique process id number). Linux also supports threads.

In addition, like Unix, Linux is a true multi-user system; a single computer can 'service' as many terminals as can be physically connected. Note that this multi-user essence is inherent in the OS. Linux "assumes" that many users will be connected to the running computer, and asks relatively little of the hardware on the user end. Basic

Table 3.2 Some useful Windows API functions.

Function Category	Function	Short Description
Graphics	BitBlt	Bitwise Blitter
	ExtFloodFill	Painting a region with a solid color
	GetDeviceCaps	Generic device capabilities routine
	Rectangle	Draw rectangle using current 'brush'
	Polygon	Draw polygon using current 'brush'
	PolyBezier	Draw Bezier Curve (spline fit array of points)
	Arc	Draw Arc
	GetGraphicsMode	Returns current graphics mode
Memory Management	CopyMemory	Copies MemAddress1 to MemAddress2; can be multibyte
	FillMemory	Fills memory
	GlobalMemoryStatus	Returns information about the global memory pool, such as virtual memory available and physical memory available
Registry Functions	RegCreateKey	Create a Registry Key
	RegOpenKey	Open a Registry Key
	RegQueryValue	Query a value in the Registry
Client/Server Process Management	CreateProcessAsUser	Spawns a process using "USER" credentials
General Process Management	CreateThread	Creates an execution thread in the current process
	GetCurrentProcessID	Returns the Process ID for a PROCESS
	GetCommandLine	Returns the command line string used to execute a running process
	GetPriorityClass	Returns the PRIORITY for the running process
	CreateJobObject	Creates a JOB in Windows
	Sleep	Suspends execution of a running PROCESS, but is stays in memory and can be reactivated
	CreateProcess	Starts a new PROCESS
	TerminateProcess	Kills a running PROCESS and returns its resources to the system

Table 3.2 continued

DLL Functions	LoadLibrary	Loads a DLL into memory, enabling calls to its functions
	LoadModule	Loads a library module into memory
	GetProcessAddress	Returns the address to use to call a function in a dll
Service Functions	CreateService	Creates a Service
	StartService	Start a Service
System Information	GetComputerName	Returns the name of the local machine
	GetUserName	Returns the current user name
	GetWindowsDirectory	Returns the Windows system directory
	GetSysColor	Returns the current system color palette
	GetSystemInfo	Returns system processing hardware information, such as number of processors
	GetSystemMetrics	Can return a LOT of system information, such as last boot type, screen dimensions, etc.
Windows Functions	CreateWindow	Create a Window

usability, not requiring a graphical user interface, is far more advanced in this setting than that in Windows.

As stated, Linux inherently accepts the concept of running executable images in the client-server model mentioned above. Windowed user interfaces in the Unix arena utilize a window-generating program on the terminal end, so that the client-server model is not disrupted. The terminal end simply must supply the graphical user interface functions, and the calls to these functions are made by the server computer over the terminal connection.

To contrast with the three user system mentioned above for MS Windows, three terminals on a Linux computer will each run applications as separate processes in separate address space, *but all on the same computer*. Networked systems can be created that operate on the Peer-to-Peer model.

Since the Linux kernel does not presuppose a graphical user interface, streamlined simple codes are relatively easy to generate. In order to generate i/o with the windowed user interface, one must add the code to do so. In this regard, the simple program for Linux is simple to generate and does not require bypassing assumed system defaults (that is, Windows 'assumes' a windowed interface, Linux does not). Demo *OS04* was 'ported' to Linux and compiled with gcc. The execution time for this version, running on the *same computer* as the Windows version, is included in Figure 3.3. While this demo possibly suggests Linux to be the 'faster' OS, the performance difference is probably more related to compiler technologies (gcc vs Visual C++).

However, even if that is the case, a faster program in gcc on Linux is a faster program, no matter the underlying cause. It is also interesting to note that the Linux version of the demo was also the smallest.

Linux represents the 'flagship' project of the so-called "Open Source" programming model. When Linux is obtained, one receives *at least* the source code for the system itself, in addition to executable images. The license to use Linux (or any software distributed under the GNU Public License) gives the user free reign to modify and distribute the code, so long as the distributions make the *original* code available. This provides an infinite variety of configuration customizations; if a system user wishes to change the way a specific routine operates, or wishes to add a function; the OS source can be changed and recompiled. Such capability is not for everyone, but advances in the OS have been made relatively rapidly from the efforts of some of the world's most expert programmers.

Linux does not store configuration information in a centralized database like the Registry used in Windows. Each program or module may have its own configuration file, so that different parts of the OS can be configured independently. These configuration files are text based and utilize a heading scheme that is easy to interpret. Newer distributions of Linux include graphical interfaces to generate and administer these configuration files, so users don't need to edit the text files directly.

Applications use configuration files similarly, so there is a single, somewhat standardized means of storing configuration information whether OS or application. Many of the system configuration files are stored in the /etc directory.

A key distinction between the Linux configuration method and that employed by Windows lies in the paradigm for module loading. In earlier MS OS's, such as MS-DOS, components were loaded at start-up by explicit mention in either the config.sys or autoexec.bat boot files. Windows dispenses with this philosophy and the installation configuration forces the loading of most components whether needed in a specific system context or not (oddly, other, seemingly commonly used components are *not* automatically loaded!). Linux, on the other hand, contains boot-up files in various run-levels that list the modules and service programs to load, and does not presuppose a 'need' for all components that may be available. By controlling the loading of modules, Linux can be streamlined to a basic OS kernel unencumbered by excess, if that is what is needed for a particular system.

To contrast, the author's installation of Windows 2000 at the time of this writing (the WINNT folder) is 953 MB, and the kernel occupies 35 MB of physical memory. Linux, on the other hand can be installed in as little as 80 MB of disk space (though rare) and can run on a machine with only 4 MB of physical RAM. Admittedly, most installations of Linux will be larger than these "minimums;" typical installations run about 400-500 MB. With regard to physical memory availability, of course, "more is better.

3.5 A BRIEF COMPARISON OF WINDOWS AND LINUX PERFORMANCE

The details of OS implementation necessarily preclude over generalization when two (or more) operating systems are compared. The software associated with 'running' a computer is very complex, and it is probably fair to say that different OS's excel at different things. However, having made that requisite disclaimer, a very brief glimpse into the empirical behavior of MS Windows and Linux is presented here; this glimpse does not pretend to test/stress all components of either OS, but rather simply presents a few clear examples of how such testing might be done.

3.5.1 Simple Numerical Procedure

Figure 3.3(a) shows the performance of three Windows versions and one Linux version of a small numerical procedure. The fairest comparison is between the Windows version of *OS04* (Windows console application) to *OS04* compiled for Linux. These two programs are essentially the same source code (the only difference being in the nature of getting and computing the execution time). The difference in the performance is more due to differences in the compilers (VC++ versus gcc), which is discussed in more detail in Chapter 4.

3.5.2 Thread/Process Creation

A more careful and detailed analysis of the inherent performance differences of these two Operating Systems was presented by Ed Bradford of IBM in his "RunTime" articles. One important function of the Operating System is to allocate resources to processes and threads and schedule the process (or thread) for execution. Dr. Bradford used very carefully constructed code to determine how many processes and threads could be created (and destroyed) on Windows 2000 Advanced Server (AS), Windows XP and Red Hat Linux 7.1 (kernel version 2.4.2) in a given time interval. The results are shown in Figure 3.4.

Another important metric in regard to process/thread creation considers how the system responds to load. Dr. Bradford also measured the rate of thread creation/destruction as a function of time; the data are shown in Figure 3.5. In this test, the Linux kernel maintained a steady rate of thread creation, whereas the Windows kernels showed a marked decline in thread creation rate the longer the test was run. This leak will cause a noticeable performance degradation in a heavily loaded system if it is not remedied. It should be noted that the OS's installed in these tests were not tuned after installation, so this comparison is between stock, 'as-distributed' kernels.

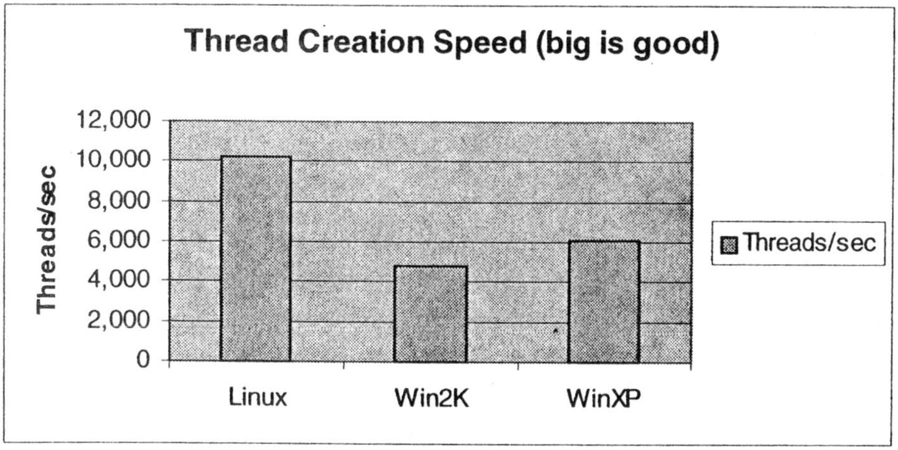

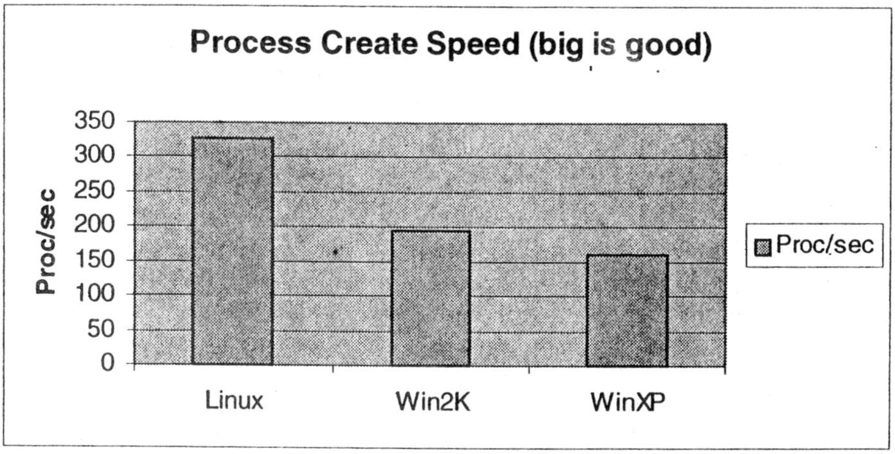

Figure 3.4: The number of processes and threads that can be created in a specified time interval for MS Windows 2000 AS, MS Windows XP and Red Hat Linux 7.2, from Bradford, Ed, "IBM Developer Works: Linux – RunTime: Managing Processes and Threads," 2002, used with permission.

3.5.3 Context Switching

One of the *most* important functions performed by a multitasking OS is context switching. A context switch is when 'control' of the system resources is changed from one process (or thread) to another. Given that CPU state, security levels, memory maps, i/o port use, file handles (and other file system objects), etc. all must be maintained *per process*, one can easily imagine just how complicated the task of context switching is. The implementation of switching contexts is very platform and OS dependent. Some systems heavily use hardware capabilities of the CPU whereas others rely more on software in the OS kernel.

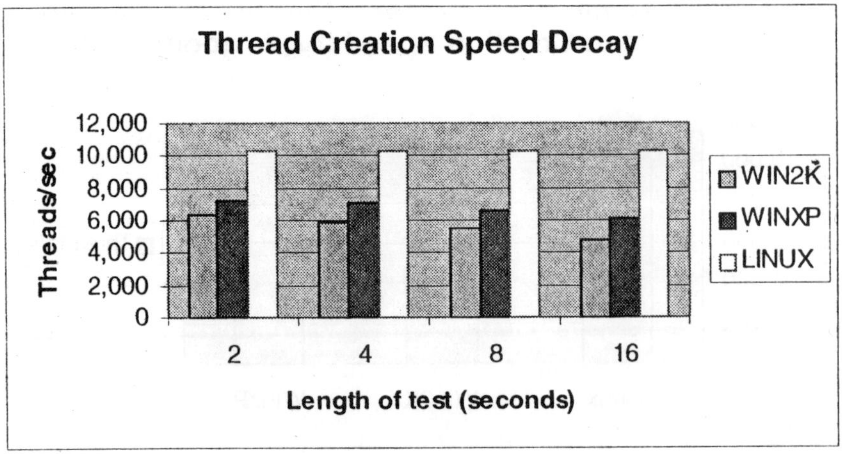

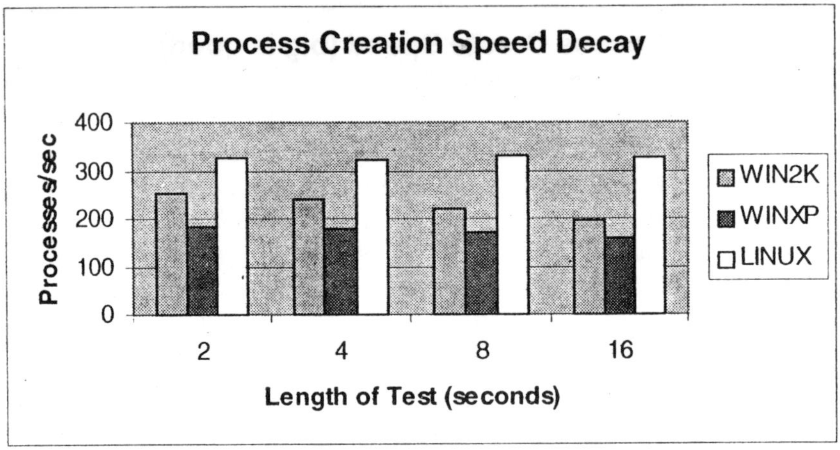

Figure 3.5: Dynamic thread creation as a function of test run time for MS Windows 2000 AS, MS Windows XP and Red Hat Linux 7.2. Linux maintains a steady thread creation rate whereas both Windows versions exhibit a performance decay as the run progresses. It should be noted that these were 'clean' installs of all OS's with no significant performance tuning. Data from Bradford, Ed, "IBM Developer Works: Linux – RunTime: Managing Processes and Threads," 2002, used with permission.

Dr. Bradford also measured the rate of context switching. This has been done by numerous authors in numerous contexts, and the result seems to be very system specific. However, partial results from Dr. Bradford's study are shown in Figure 3.6(a). This data, from a program that uses a pipe to pass a single byte from thread to thread, seems to suggest Linux is far faster at switching contexts than either version of Windows. However, this does not tell the whole story. Linux is known to implement pipes faster than Windows, so the choice of the *token* passed between threads influences the result in Figure 3.6(a). Bradford also presented data using the *optimal* token for each OS (critical section for Windows and mutex for Linux), thus optimizing the token passing overhead for each OS. When this was taken into account, the *net* context switch time for both Windows versions was about half that of Linux; data for this optimized test are shown in Figure 3.6(b). As an added consideration, the threads merely got the CPU and released it in these 'sterile' tests. Actual working threads will be part of a future study. Finally, Bradford did present data dealing with context switching when multiple threads were concurrent. However, for brevity, this is omitted here.

From the preceding paragraphs, it is clear that OS or platform comparisons are technically difficult to do in a controlled way. Further, it is clear that Linux performs better than Windows in regard to process/thread creation, but Windows is faster at switching between the processes/threads once they are created. However, though it should be noted that the process scheduling code is one of the major improvements in the Linux kernel version 2.6.2, so the data in Figure 3.6 do not represent the state of the art for Linux. In addition, as noted, Bradford used mostly 'untuned' OS's in his tests, and kernel tuning may result in different test results. In this regard, being open source and highly configurable, Linux would probably be considered the more flexible option.

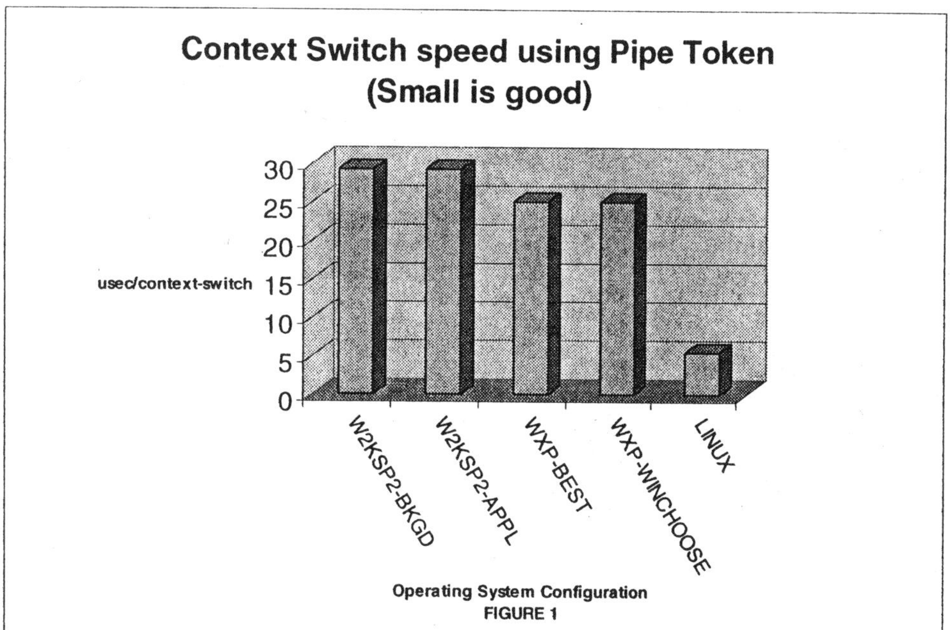

FIGURE 1

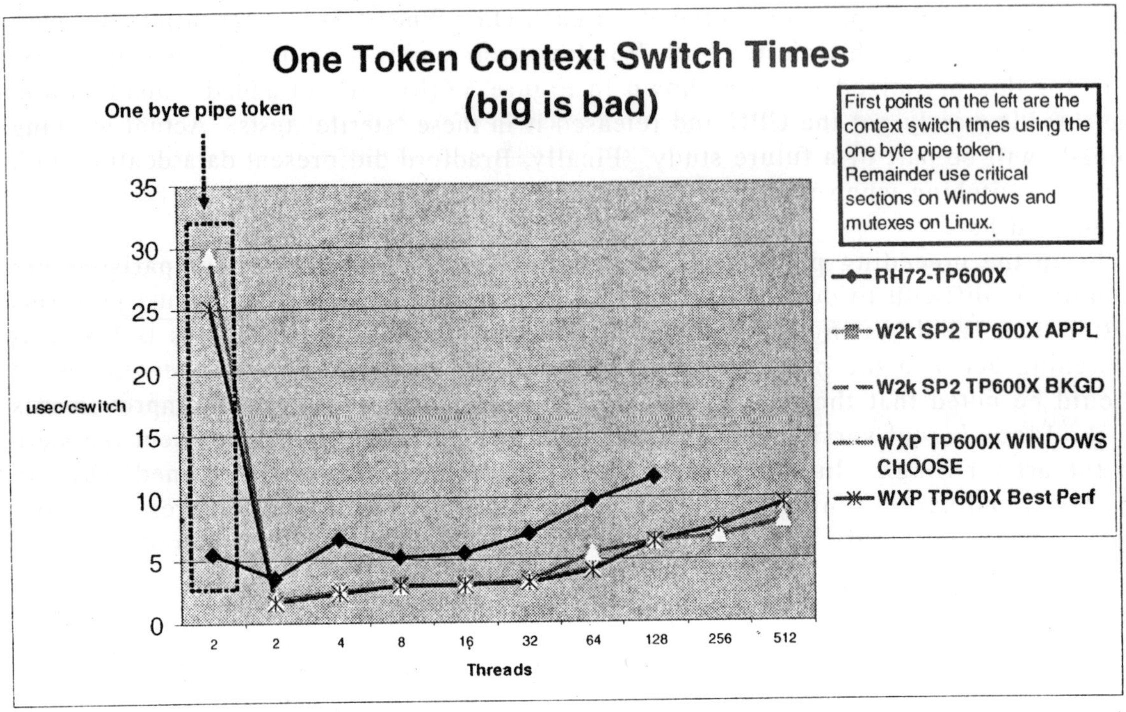

Figure 3.6: Raw context switch times for MS Windows 2000 AS, MS Windows XP and Red Hat Linux. The data represent two different tune settings for the Windows versions are shown. Figure 3.6(a) shows raw context switch times when a single byte token is passed between threads via a pipe. The token overhead is assumed negligible in these data, which is not the case for the Windows kernels. Figure 3.6(b) shows raw context switch times as a function of number of threads when the OS dependent *optimum* token passing is used (critical section for Windows, mutex for Linux). Data from Bradford, Ed, "IBM Developer Works: Linux – RunTime: Context Switching Part I," 2002, used with permission.

Chapter 4: Compiler Considerations

Not all compilers are created equally. This may seem like an obvious statement, but it is mentioned here for emphasis. In particular, high-level source code written for older compilers, hardware and operating systems does not take advantage of modern technology. For example, running code written or compiled for 16-bit DOS on a 2000 MHz 32 bit machine with 512 MB of RAM is tremendously inefficient.

Even modern compilers vary considerably in the features, ease of use and built-in optimizations that are available. In this Chapter, some of the issues important to selecting and using compilers are examined. To present these issues, some compiler optimizations and the evolution of several modern compilers will be discussed. Not all compilers can be addressed, so in keeping with the model of previous chapters, only a sample is presented. However, as with previous material, the *method* of selecting a compiler (or evaluating its output) is what is being demonstrated. Numerous resources regarding programming languages and compilers are listed in Appendix A.

4.1 INTERPRETERS VS. COMPILERS

BASIC is an example of a language that has grown over the years from a strictly interpreted language to a compiled one. Interpreted languages have the code parsed and compiled "on the fly" during execution, rather than compiled and linked to an executable. The advantages of interpreted languages include easy interactivity during development, but this is dulled somewhat by the Interactive Development Environments (IDE) of many modern compilers. On the disadvantage side, interpreted languages require the interpreter be present (that is, no 'stand-alone' executable is created) and, importantly, they are *slow*.

The slowness of interpreted code arises from two key run-time characteristics. First, the code must be parsed and compiled in a single pass; the interpreter cannot use information it has not yet encountered. Second, the parsing and compiling takes time. Interpreted languages may be useful for algorithm testing and other development, but should never be relied upon when high performance is required.

When code is compiled, the compiler software may use very advanced algorithms to determine optimum object code. This is helped by multi-pass compilation in which the compiler gathers key data about the code in an early pass and refines the object code with later passes. Not only is the code not parsed and compiled at run-time; but also the object code is likely more efficient than that from interpretation. Therefore, all performance applications should be compiled.

Java is an example of a *modern* language that is generally interpreted. While Java is generally 'compiled' to bytecodes, at run time, the bytecodes are interpreted. This approach eliminates *some* of the interpretation problems mentioned above, but the language remains fundamentally interpreted; the Java language specification precludes actual compiling. However, so-called "Just In Time" (JIT) compilation tools have evolved that formally compile the bytecodes 'on the fly' at run time. These JIT 'compilers' do improve performance of Java when compared to interpreted Java. However, the JIT compiler cannot perform multi-pass optimizations.

4.2 COMPILER OPTIMIZATIONS

In Chapter 2, two different methods of looping in ASSEMBLY Language were demonstrated to illustrate that different code logic might produce different execution efficiencies (**Listings 2.4** and **2.8**). This applies to high-level languages as well. However, the issue is further complicated by the fact that compiler optimizations may result in different object code from the same source code. So, besides 'knowing your compiler,' how does this relate to programming in a high level language, such as C? There are at least two ways to write code for a simple for loop in C, and they can be compiled either using compiler optimizations or not. **Listings 4.1** and **4.2** show the MS Visual C++ output for a simple loop for two such compilations.

Listing 4.1 Non-optimized compiler output for a C for loop with increasing loop counter

```
for(c=1; c<=100; ++c){

      mov   DWORD PTR _c$[ebp], 1
      jmp   SHORT $L583
$L584:
      mov   eax, DWORD PTR _c$[ebp]
      add   eax, 1
      mov   DWORD PTR _c$[ebp], eax
$L583:
      cmp   DWORD PTR _c$[ebp], 100
      jg    SHORT $L585

      }
```

```
      jmp   SHORT $L584
$L585:
      {next line of code}
```

For the loop indicated, that is looping 1-100, this code has

1	MOV	mem, imm	for a total of	1 Pentium cycle
101	JMP	short	for a total of	101 Pentium cycles
101	MOV	reg, mem	for a total of	101 Pentium cycles
100	MOV	mem, reg	for a total of	100 Pentium cycles
100	ADD	reg, imm	for a total of	100 Pentium cycles
101	CMP	mem, imm	for a total of	101 Pentium cycles
1	JG	short (jump)	for a total of	1 Pentium cycle
100	JG	short (no jump)	for a total of	100 Pentium cycles

This code bears some resemblance to the code in **Listing 2.8**, so its efficiency is clearly suspect. In fact, **Listing 4.1**, requires 605 cycles to execute on a Pentium.

With the compiler set to optimize for speed, the same source code compiles to the ASSEMBLY instructions shown in **Listing 4.2**.

Listing 4.2 Optimized compiler output for a C loop with increasing loop counter

```
for(c=1; c<=100; ++c){
      mov   eax, 1
$L583:

      inc   eax
      cmp   eax, 100              ; 00000064H
      jle   SHORT $L583

}
```

This code is comprised of

1	MOV	reg, imm	for a total of	1 Pentium cycle
100	INC	reg	for a total of	100 Pentium cycles
100	CMP	reg, imm	for a total of	100 Pentium cycles
99	JLE	short (jump)	for a total of	100 Pentium cycles
1	JLE	short (no jump)	for a total of	1 Pentium cycle

Completing in 302 cycles, this ‘version’ runs just over twice as fast as that shown in **Listing 4.1**. The code in **Listing 4.2** is analogous to the code in **Listing 2.4**.

The other way to write the loop in C is

```
for(c=100; c>=1; --c){

}
```

Compiler output for the non-optimized version was only a little different. For the optimized version, the key change was substitution for `DEC EAX` in place of the `INC EAX`, and the change of `CMP` to `TEST`. Neither of these changes affects the overall latency of the routine, however, so the timing of the optimized code is the same whether the loop counts from 1-100 or from 100-1.

In the preceding discussion, the very general 'optimize for speed' compiler switch was used. In the remaining parts of this section, some specific compiler optimizations are briefly described. Not all compilers offer each option, and certainly, other compiler optimization options exist.

4.2.1 Aliasing

The 'Assume No Aliasing' option allows the compiler to utilize registers and other optimizations when it is known that each memory location has a *unique* symbolic name. That is, it must be known that a single memory location is referenced by only one variable (or function) name. An example of aliasing is having a regular variable and a pointer variable both representing the same address in memory. The advantage of aliasing to the programmer is convenience in defining variables names (and pointers) as needed. The key disadvantage of aliasing is that, at the ASSEMBLY Language level, the pointers must be reloaded whenever an alias is encountered. This precludes the consistent storage of address references in cpu registers and is, in a sense, a needless moving of data.

Some compilers allow the assumption of No Aliasing within function calls, but allow the use of aliases across function calls. This is a middle ground between no aliasing and full aliasing. Some overhead in fundamental addressing is required, but only when a function call is complete; that is, the aliases must be 'reset' after function calls. Note that this is not an optimization at all for code that calls many short functions, such as that illustrated in Listing 6.2.

By assuming "No Aliasing," the compiler can generate more efficient object code in regard to memory references. In particular, loop parameters can be streamlined and efficient register use results when memory references are not aliased. This style of programming, at the high-level source code level, requires careful planning and attention to how data is referenced. This planning applies to function call stack manipulations (local variables) as well as the general, overall layout of data in memory.

4.2.2 Array Bounds Checking

Some languages provide the programmer with strong checking of array elements. That is, the programmer defines an array length, and at run-time, code exists to determine if a reference is within this length. Other languages, such as C/C++, do not include this failsafe. With this feature, an array reference outside the limits of the defined array generates a run-time error; though the program may crash, this is a much softer failure than reading (or changing) the 'wrong' memory address (which may produce very unpredictable results).

This constant checking of array references consumes CPU cycles otherwise spent on useful work, however. If the source code is properly constructed, and bug tested, there should be no need for this safety net in working, production code. Compilers of languages that generally provide such tests often allow disabling the tests. This compiler switch results in significantly faster code in regard to array references (reading or writing array elements).

4.2.3 Numeric Overflow Checking

Some compilers insert numeric bounds checking by default. This code checks the values of numeric data types, such as integers and floating point data, to ensure the value lies within the limits of the declared type. Is this necessary for general code? Imagine a loop counter that varies from 1 to 100. Cycles used to constantly test this value are wasted, and many compilers allow these checks to be disabled. Indeed, this places a greater burden on the programmer to structure the code in a fashion that does not allow (or at least catches) potential overflows, or the program may not behave as expected.

4.2.4 Unrounded Floating Point Operations

The way data is moved to/from the x87 registers results in some uniform rounding of the data. Recall from Chapter 2 that the x87 registers are 80 bit, so contain more precision than the operands, whether single precision (32 bit) or double precision (64 bit). By including code to keep the rounding of 'intermediate' results uniform, the compiler ensures that two values are equal when compared. However, this also results in excessive moving of data from registers to memory, which wastes cycles. These cycles can be recovered by disabling the compiler feature that forces the rounding of floating point values. This should be used with caution, however, since this may result in comparisons failing when the values 'appear' equal. That is, the compared values are equal to 32 or 64 bits, but are different to 80 bit precision. Without the safeguards in rounding, the extended precision can result in comparison of the 80 bit data, which may result in a comparison failing.

4.2.5 FDIV Bug Checking

The original Pentium processor contained a hardware bug that resulted in an error when a floating-point division was performed. This bug was fixed while the Pentium was in production, so there are probably very few current computers using the faulty chip. Some compilers are designed to be 'downwardly compatible' to the Pentium by default and include code to correct this error. As stated previously, software is generally slower than hardware, so performing floating-point division (or correcting) in software is slower than utilizing the hardware processor. If it is known that the software will not run on a Pentium containing the FDIV bug, disabling the bug checks in the compiler will produce faster division code. This is an example of compilers producing code for legacy hardware negatively impacting performance.

4.2.6 Inline and Intrinsic Functions

As will be discussed in detail in Chapter 6, calling functions and procedures produces object code necessary to administer the function call. This code must execute for every function call, and, in a way, represents wasted cpu cycles. Many compilers provide options to minimize the impact of function calling, such as generating intrinsic or inline functions. It should be noted, however, that this compiler option does not *always* insert explicit code; there generally exists some logic in the compiler code to determine which functions to 'inline' and which to not. Often, this is based on the size of the code in the function itself.

One way that function calling is inefficient is that a 'new' stack must be generated for the function; each function gets its own stack space (within the stack space of the calling program or function) structured as a so-called 'frame pointer.' In addition to reducing function calls themselves, some compilers allow the elimination of frame pointer creation for those function calls that are made.

4.2.7 Common Sub Expressions

Sometimes a function or loop will contain code that does not need to be computed where it appears in the source code. Consider for example, the code in **Listing 4.3**.

Listing 4.3 A Small Loop Containing a Common Expression

```
for(i=1;  i<=1000;  ++i){

        b  =  10.23;
        c[i]  =  b  *  i;

}
```

In the code in **Listing 4.3**, the assignment of b inside the loop is redundant;

b = 10.23 is a *common sub expression* (cse) to all iterations of the loop. Some compilers contain optimization algorithms to migrate the *b* assignment from inside the loop to outside the loop. The compiled object code is as if the source code were written as shown in **Listing 4.4**.

Listing 4.4 Small Loop with Common Expression Outside the Loop

```
b = 10.43;

for(i=1; i<=1000; ++i){

      c[i] = b * i;

}
```

This type of optimization can be applied by the compiler to code within functions as well, which can result in reduced time spent in the function.

4.3 USING PROGRAMS COMPILED WITH OLD COMPILERS

An execution time comparison (which is actually Demo *RISC01* from Chapter 8) for executables created with two compiler versions (a modern, Win32 one and an older, DOS 16 one) is shown in Figure 4.1. The times are for the *same source code and the same computer running MS Windows 2000*. This demonstration clearly shows that the compiler definitely impacts code performance, especially when mismatches between compiler technology and run time hardware/OS exist.

4.4 C++ AND SIMILAR COMPILERS

There are many C++ compiler products on the market. This section includes a *very brief* overview of several such compilers. Included here for discussion are MS Visual C++, g++ and VectorC, as well as Visual C#. Additional C/C++ compilers (free and commercial) are listed in Appendix A.

4.4.1 MS Visual C++

Arguably, the most used C++ compiler for Windows based software is the Microsoft product Visual C++ (VC++). VC++ includes an Integrated Development Environment (IDE) that allows code editing, compiling, linking and running, as well as the use of

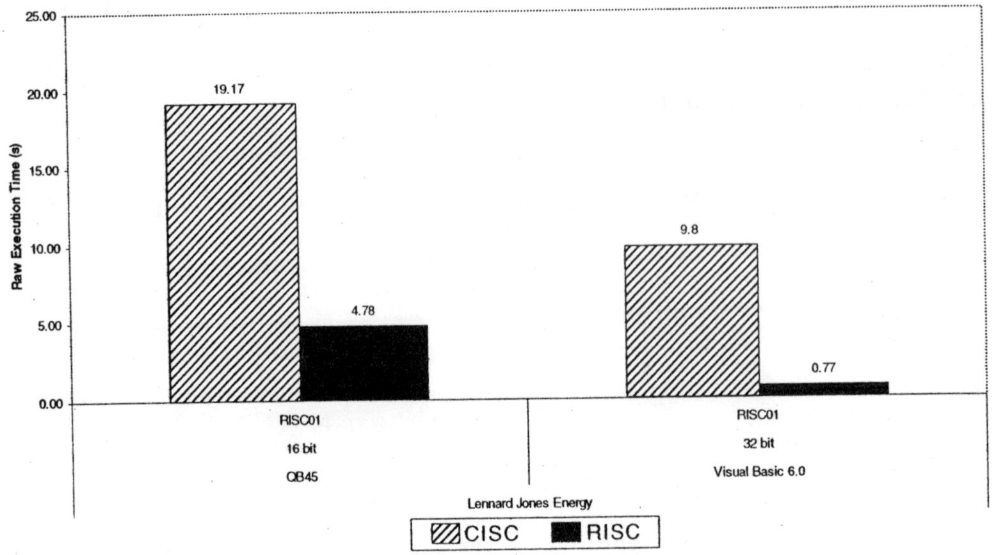

Figure 4.1: Demo *RISC01* (from Chapter 8) raw execution times for code compiled on two Basic compilers (16 bit Quick Basic 4.5 and 32 bit Visual Basic 6.0). Both versions were run on the same test computer running Windows 2000. This figure illustrates the potential performance cost to using code compiled with older compilers even on modern computers with modern operating systems.

several tools (such as profilers and debuggers) used from within the IDE. VC++ includes many basic and advanced optimization options, such as Inline Functions, as shown in Table 4.1. In addition, many types of applications are easily created, such as dynamic link libraries, executable applications and Component Object Model (COM) components.

VC++ is an object oriented environment and allows the user to quickly create basic Windows programs via the use of Wizards. That is, much of the process of creating and registering Windows classes with Windows can be 'automated' in VC++, and integration with the Windows API is very tight. Using Microsoft Foundation Classes (MFC), many useful classes are available to the programmer, but this includes large run time overhead in the program. Extremely large projects are generally compiled from the command line and can be managed with 'makefiles.'

Interestingly, MS Visual C++ compiler optimizations may be a little too aggressive. The "Optimize for Speed Option," /O2, implies 'Global Optimizations, /Og. Global Optimizations causes the compiler to scan the source code for *common sub expressions* (Section 4.2.7) that do not change, and these are 'hard coded.'

This is a powerful feature when it works, but the compiler can 'miss' code that changed. For example, consider Demo *OS04* from Chapter 3, a console application that computes a random distance and Lennard Jones energy for that distance (this comment also applies to *OS02* and *OS03*, the windowed versions of *OS04*). The demo

Table 4.1: A brief comparison of optimizations available for several compilers.

Optimization	VC++6.0	VectorC	g++
Address Space	32 bit	32 bit	32 bit
Aliasing	+	+	+
Alignment	+	+	+
Remove Array Bounds Checking	0	0	0
Remove Numeric Overflow Checking	0	0	0
Unrounded Floating Point Operations	+	+	+
FDIV Bug Fix	+	-	-
Inline Functions	+	+	+
Loop Unrolling	-	+	+
Migrate Common Sub Expressions	+	+	+
Advanced Instruction Sets (Chip Specific)	-	+	+

\+ This optimization option is present

– This optimization option is not present

0 does not apply; the language itself does not include these checks

uses the hand-rolled pseudo random number generator from Chapter 6 (since the call to rand() can be very slow). In a recursive equation, the code computes *xk_1* using *xk* then assigns *xk* = *xk_1*. The Global Optimization algorithm 'thinks' *xk* is not changing, so the code is 'removed' from the loop (that is, the cse, is migrated). The compiler then generates no code for the loop, and executes the loop in "0" seconds! This behavior was noticed for many of the samples compiled for this book, and that is why many of the demo programs don't use the compiler optimizations. The Microsoft Developer's Network contains numerous articles outlining similar failures of the cse optimizer in VC++; two of these are listed in the Bibliography. This should be remembered when profiling code compiled with compiler optimizations enabled; the compiler may not generate object for code blocks it 'thinks' are unnecessary. In other words, this illustrates that compiler optimizations are handy, but not a substitute for careful programming..

4.4.2 MS Visual C#

C# is a C language variant that has been designed from the ground-up to favor low development time and easy integration with web based applications. Modeled after MS Visual Basic, C# is heavily object oriented; in fact, all objects are COM objects, an approach that reduces work for the programmer, but dramatically increases compiled code overhead. In other words, this language intrinsically uses a very high degree of data and function abstraction.

C# in a sense relaxes some of aspects of C/C++ that make C a 'programmer's language.' For example, C/C++ does not support automatic variable initialization, whereas C# does. C# is a new technology, but it appears that this system is entirely unsuitable for true high performance computational code.

4.4.3 g++

g++ is the GNU C++ compiler shipped with many distributions of Linux. As shown in Table 4.1, g++ includes more robust optimization options to the programmer than the Microsoft products. For example, g++ includes many machine specific optimizations. This is clearly an example of the different application deployment paradigm used for Linux compared to MS Windows. For Windows, compiled and linked *binaries* are deployed so machine specific optimizations are not generally practical; on the other hand, with Linux, source code is often deployed and *compiled by the user*, with machine specific optimizations, on the user's machine directly. The g++ compiler is typically invoked via the GNU Compilier Collection (gcc) 'supercompiler.' Further, unlike VC++, g++ supports code generation for a wide variety of architectures in addition to x86 processors.

g++ does not itself include an IDE, though third party products are available (such as KDevelop for the KDE environment). Without an IDE, one edits the source with a text editor, and then compiles and links the code with gcc; any use of a debugger or profiler (or other tools) must be done 'manually.' g++ does support the creation of dynamic libraries and other application types, as well as standard executable applications. Most applications are 'console' applications, unless created and compiled with toolkit GUI libraries. Large projects can be managed with 'makefiles.'

The abbreviated summary in Table 4.1 does not tell the whole story in regard to g++ optimizations. For example, not only does g++ allow the user to select "Alignment," but the types of alignment can be selected as well, such as aligning functions, jumps, loops and labels. Also, note that g++ allows the compiler to unroll small loops (see Chapter 7), which is an optimization not provided by the Microsoft compilers.

Since gcc (and therefore g++) has been developed for 'any' system running any 'variety' of Unix, there are generalities in the code created by this compiler. Specifically, the compiler does not take advantage of architecture specific hardware optimizations by default. Therefore, one of the most important command line switches for gcc is the -march switch (in addition to the -O3 switch), which sets the architecture for which the code is being compiled. For example, the code in the Linux version of Demo *OS04* presented in Chapter 3 was compiled using -march=athlon for the test computer. The execution time for *OS04* without this switch (that is, using the 'default' architecture behavior) required 5.90 seconds; a factor of 3.5 performance improvement was gained by specifying the architecture. The reason the -march switch is so important relates to material presented in Section 2.5. Without -march properly set, the code is not properly scheduled for the run time CPU. The command line used to compile most of the Linux demos for this book was:

```
gcc -Wall -O2 -o <output filename> <source filename>
-lm -march=athlon.
```

Of course, readers compiling their own versions from the CD-Rom Linux source code should compile with the appropriate -march= architecture switch (or use the makeall.sh script to compile all the programs).

4.4.4 VectorC

VectorC is a high performance straight-C compiler developed by CodePlay. The compiler includes intrinsic capability to produce Single Instruction, Multiple Data code without the need to write inline ASSEMBLY in the source code. In addition, the compiler will generate 'hints' for the programmer to set certain compiler switches for certain cases. These include situations for which it is unsafe for the compiler to 'assume' an optimization, such as memory alignment. This is a powerful compromise to having the compiler actually performing potentially risky optimizations.

VectorC is currently released in several versions, including one for Win32. The Win32 release can function as a stand-alone compiler or can be used as an 'add-in' for MS Visual Studio. However, at the time of this writing, the compiler is 'straight C,' but CodePlay does indicate a C++ release is being developed. In addition, there is a Linux version being developed, but has not been released. A demo version of VectorC is on the CD-Rom that accompanies this book.

4.4.5 KAI C++

Though it has not been sold since April 2002, KAI C++ is mentioned here to present a simple contrast that fits nicely into the run-time performance versus development time that is an underlying theme throughout this book. KAI C++ was developed to enhance 'programmer productivity,' specifically in regard to cross platform portability. In other words, this compiler's focus was on strict portability of ISO C++ code and therefore was less than optimal for use with specific hardware optimizations.

Chapter 5: Data Management

The manner of memory allocation for data handling is crucial to high performance programming. Some high-level languages, such as C and Pascal, use a paradigm that requires memory be properly allocated *before* code accessing that memory is executed. This forces the programmer to plan memory allocation, but does not necessarily ensure efficiency.

Other languages, such as C#, BASIC and PERL, do not use the explicit allocation paradigm. This means that memory allocation can be done 'on the fly,' implicit in a memory access statement (such as a variable reference). Since the memory allocation is not specified by the programmer, the most efficient use of memory cannot occur.

Data typing has significant consequences for object code generation by the compiler. As a simple example, consider the statement $a = a + 1$, which can be compiled to an `INC` op-code only if a is an integer type. If, on the other hand, a references a memory location holding floating-point data, the `INC` op-code cannot be used. One thing to remember is that while in algebra variables are abstractions of quantities, in programming the situation is different. The variable name is an abstraction of a *reference to memory*. How that reference is abstracted obviously influences the compiled op-codes.

In Chapter 5, some general considerations of data management are presented. Proper data management provides the foundation for other optimizations discussed, since poorly structured data impacts any algorithm that manipulates that data. Further development of these ideas is presented in subsequent chapters.

5.1 IMPLICIT DECLARATION AND THE VARIANT PROBLEM

This Chapter begins with addressing implicit declaration. The information in this section does not technically apply to languages that require explicit allocation, but serves to illustrate perhaps *why* explicit allocation is required. By not requiring explicit allocation, languages such as C# allow programmers to begin writing code quickly and devote relatively little time to memory management. Implicit declaration occurs when an undeclared variable is used in a programmatic statement, such as `x = 20.3` (where, in this example, this is the first occurrence of x in the code). Problems from using implicit declarations include hard-to-track bugs, excessive memory usage and very slow code.

Buggy and hard to debug code may result from using implicit declaration. In part, this is due to the compiler not catching typographical errors in declaration statements. The author recalls a commercial high school scheduling program (written in BASIC) running on an Apple IIE in 1983 that scheduled everyone in the school for the same class because a variable reference to QQ was confused with the variable Q. As a specific example, consider **Listing 5.1**.

Listing 5.1 **BASIC Code showing a typo in variable declaration**

```
' intended to type Energy, not Enrgy!!
Enrgy = 10.6

{a bunch of code}

' Energy = 0, not 10.6!
Needed_Value = Energy * 2
```

In **Listing 5.1**, the variable `NeededValue` *always* evaluates to zero, since `Energy` is zero. `Energy` is not initialized, so the compiler implicitly declares `Energy=0` when the last statement is encountered. This kind of typographical bug *can be* very difficult to identify, since quick glances at the source code result in `Enrgy` being misread as Energy.

Slow code can result from implicit declaration since generally the compiler will allocate all implicitly declared variables as some generic type (for example, in Visual Basic, this type is a Variant). Not only can this be wasteful of memory (Variants always use at least 16 bytes), but also forces the compiler to generate type conversion code. Data type conversions *must* occur before the data can actually be used, and such type conversions are performance killers.

Why must a type conversion occur? Consider for example the language level sin() function. This function takes a double precision value (the angle in radians) as an argument, not a Variant. A prototype for sin() may be written

```
Function Sub sin(byref angle as DOUBLE) as DOUBLE
```

If the angle passed to the sine function is a VARIANT, the compiler must insert code to convert the angle to a DOUBLE before the function call (to sin()) can be made.

This also illustrates another flaw of the implicit allocation: no 'Type Mismatch' error occurs. In contrast, the compiler inserts the type conversion code. (Said another way, the compiler 'senses' that the programmer is not proficient enough to pass properly typed variables to functions, so always has the ability to add the conversions). Similar op-codes result when an integer is passed to `sin()`, and again, no 'Type Mismatch' error occurs. The implied VARIANT declaration performance is contrasted with explicit properly typed declaration in Figure 5.1 (Demo *Data01*).

Also demonstrated by *Data01* is the general problem with type mismatches that can occur when variables are not properly declared. Some languages, such as Pascal, use a 'strong typing' paradigm and do not allow type conversions. Other languages, such as C/C++ do allow type conversions (called 'cast' in C parlance). However, these conversions can be very slow, especially when converting between floating point and integer types. The C standard library ftoi() function, for example, is known to be slow and faster float to integer implementations have been developed.

Many languages that allow implicit declaration do provide directives to combat part of the implicit declaration problem. For example, Visual Basic utilizes "Option Explicit" and Perl implements 'use strict.' The use of these directives reduces the 'buggy' problem with implicit declaration. For example, with `Option Explicit`, **Listing 5.1** causes a *compile-time* error advising the programmer that the variable `Energy` is being used, but is not declared. Likewise, if `Energy` is declared, but Enrgy = 10.6 is the initialization line, a compiler error also occurs. Option Explicit, therefore, can be used to catch *some* variable name typographical errors that the

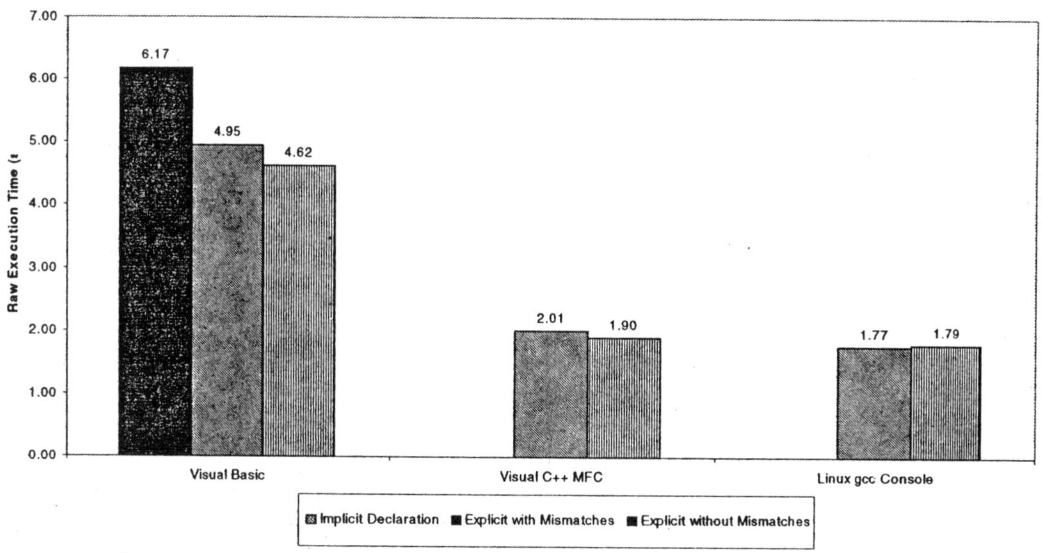

Figure 5.1: Demo *Data01* raw execution times on the test computer for 10,000,000 iterations, which demonstrates the use of simple variables. BASIC allows implicit declaration (which defaults to type VARIANT), and this is compared to explicit declaration with matched and mismatched types. C/C++ does not allow implicit declaration, so that function is not shown for the C++ example.

compiler would otherwise ignore.

The second problem with implicit declaration is not necessarily eliminated by an Option Explicit type directive. For example, **Listing 5.2** causes no compile-time error (or runtime error, for that matter) even though three type mismatches occur. The first type mismatch is subtle: $\pi/4$ evaluates as a Double, but is assigned to a Single. The second is that the value passed to the sin() is a single not a double. The third is that the sin() function returns a double, which is assigned to a single variable in this code example.

Listing 5.2 Single precision value passed to function expecting double causes no error in BASIC

```
' first line in module, outside
' any functions
Option Explicit

Dim Angle as Single
Const PI = 3.1415926536
Dim NeededValue as Single

Angle = PI/4

NeededValue = sin(angle)
```

Since the type mismatches occur, the compiler again inserts type conversion code. The result is slower code than if `Angle` and `NeededValue` were declared as Double. In a language such as C, this type of code generates an error (or warning) at compile time.

5.2 ELIMINATING TYPE CONVERSIONS

The data type conversion problem exists in any language that does not enforce strong typing. Languages that require explicit allocation (C/C++) will not allow the programmer to 'perform' these conversions implicitly, but this does not insure that programmers allocate memory references in the most efficient way. As mentioned above, data type conversions increase execution time. These conversions appear consistently in scientific software since the Scientist Programmer (as defined in Ch. 1) typically programs either whatever data type is convenient or uses the maximum precision at all steps in the belief this will make the calculation more precise.

5.2.1 Type Matching to Function Calls

One of the problems with improper typing, especially 'always' using doubles is that many of the built-in functions operate on or return specific data types. For example, the GetTickCount() function in the Windows API returns a float, so code like

```
int myvariable;

myvariable = (int)GetTickCount();
```

is inherently slow due to the type conversion. Poor choice in declarations results in CPU cycles being wasted on type conversions; sometimes, if the code is not carefully planned, these conversions occur back-and-forth between types. If some conversion cannot be avoided, the effect can be minimized by carefully profiling the algorithm. In short, the conversions should not be done more than is necessary, and are generally avoided inside large inner loops. Another optimization may be to hand-code a function acting on (or returning) the type of interest rather than always relying on built-in functions that force the programmer to use type conversions; there will, however, be cases where the type conversions are quicker with the built-in functions, especially when the built-in function is 'close' to hardware.

5.2.2 Loop Counters

It is well known that the integer data type is generally the fastest for loop counters. This is especially true when the counter is incremented by 1, so that the integer `INC` reg op-code results in the compiled code. However, this general idea assumes that the counter is not used inside the loop where a floating-point value is needed. In this case, the type conversion cost may outweigh the cost of using a floating-point counter.

To illustrate this, Demo *Data02* uses the sin() function to compute the sine of angles 0–360 degrees in 0.01 degree increments. To program this routine using an int loop counter, code such as **Listing 5.3** is used.

Listing 5.3 Floating point arguments with an int loop counter

```
// conversion factor for
// degrees to radians,
// divided by 100
double cDeg2Rad = PI/18000;

int Counter;
// holder for the computed sin
double dummy;
```

```
// int cannot step 0.01,
// so actual 'degree' * 100
for(Counter=0; Counter<=36000; ++Counter){

      // Counter must be converted to
      // double precision radians
      dummy = sin((double)Counter * cDeg2Rad);

}
```

The code in **Listing 5.3** requires an integer to double precision floating-point conversion in the sin() argument. Consider the contrasting code in **Listing 5.4** that involves no type conversion.

Listing 5.4 Floating-point arguments with floating point loop counter

```
// degrees to radians
double Deg2Rad = PI/180;

double x;
double dummy;

// x is a Double Precision counter
for(x=0; x<=360; x+=0.01){

      // though argument is already a double
      // it must be converted to radians
      dummy = sin(x * Deg2Rad)

}
```

Execution times comparing **Listings 5.3** and **5.4** are shown in Figure 5.2. The double precision loop counter in is slower for MS Windows but faster for Linux than the integer loop counter with conversion code. It should be noted that there may be cases which has the integer loop counter slower for Windows as well, so *always* using integer counters without thought to how it is used in the algorithm may produce relatively low performance code.

Another question is the choice of units for the counter. In this example, we want the independent variable in degrees, but the sine function takes radians as an argument. Therefore, the loop could be constructed in radians rather than degrees. This may save the compute cycles involved in converting degrees to radians inside the loop. This

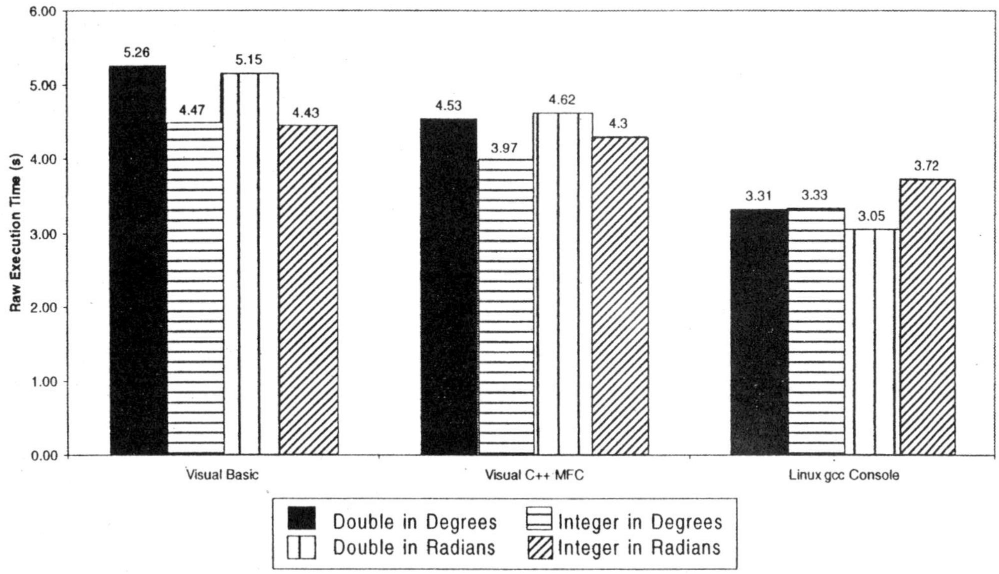

Figure 5.2: Demo *Data02* raw execution times on the test computer for 1,000 iterations. In this demo, sin(x) is computed from 0 to 360 degrees, in 0.01 degree increments. There are four ways to code this computation: using a double precision loop counter in degrees, a long integer counter in degrees, a double loop counter in radians or a long integer counter in radians.

requires transformation of three parameters from degrees to radians (either by the programmer or, if in code, *outside* the loop):

Loop Start: 0 degrees = 0 radians
Loop Stop: 360 degrees = 2π radians
Loop Step: 0.01 degrees = $2\pi/18000$ radians

With these transformations, **Listing 5.4** becomes

Listing 5.5 Floating-point loop counter without unit conversion

```
double Arg1;
double dummy;

// cDeg2Rad defined as in Listing 5.3
for(Arg1=0; Arg1<=Pix2; Arg1+=cDeg2Rad){

      dummy = sin(Arg1);

}
```

No floating-point math, other than the desired sin() function, is performed in **Listing 5.5**. In other words, **Listing 5.5** is very efficient since there are no type conversions and virtually all cpu effort goes toward useful work.

For completeness, the 'radian space' calculation using a int loop counter is also presented. This code is shown in **Listing 5.6**.

Listing 5.6 INTEGER loop counter without degree to radian unit conversion

```
// declarations as in Listing 5.3

// counter and step are longs
for(Counter=0; Counter<=62831853);Counter+=1745){

      // sin() gets a double as argument
      dummy = Sin((double)Counter * cRadStep);

}
```

The loop boundaries and step size in **Listing 5.6** were chosen so that the step size is properly cDeg2Rad. Execution times for **Listing 5.5** and **Listing 5.6** are also shown in Figure 5.2 (Demo *Data02*). Again, for *this* demo, for MS Windows, the integer counter was faster but the floating-point counter produced the faster code with gcc in Linux. Of the four methods tested, the fastest method may be language and particular algorithm dependent, so the programmer should test the algorithm using different methods for loop counting.

Why might **Listing 5.6** be slower than **Listing 5.3**, even with the absence of the floating-point unit conversion (as it is for both Windows and Linux)? This can be explained by the step size in **Listing 5.6**. That is, when the code requires the next value for the loop counter, the step size must be fetched from memory in an `ADD reg, mem` op-code (or a fetch to register followed by `ADD reg, reg`). In **Listing 5.3**, on the other hand, the step size is 1 so that the compiled code is simply an INC reg.

The reader may be wondering why an example was presented that confirms (for Windows programs) the conventional wisdom concerning loop counters. The point is that the result shown in Figure 5.2 is not universal. For the *same* demo compiled with gcc for Linux (results are also shown in Figure 5.2), **Listing 5.5** was faster. This may be due to more aggressive compiler optimizations. In any case, the point is that using float or double loop variables *may* be faster in some circumstances. If the code inside the loop forces type conversions of the loop counter, one probably should consider using a non-integer loop variable.

5.3 IMMEDIATES, CONSTANTS AND VARIABLES

The way the data is 'stored' in the compiled program impacts run-time performance. Data stored in memory must be retrieved from memory before use, which may well be a performance bottleneck (though it may be 'prefetched' by modern processors). On the other hand, immediate data, whose values are 'hard-coded' into the program, do not require such transfer; the values can be directly assigned to registers when needed.

If a value may change during execution, the data is obviously a variable and must be stored in memory. However, if a needed value will not change (such as universal constants, multiplicative factors, etc), the value can be programmed as a 'constant' or even entered as immediate data.. Using constants, the value is stored in memory, so it must be fetched when (or before) it is needed.

The use of immediate data in source code is considered somewhat cumbersome. The numbers often do not convey as much information as symbolic references. With some languages, the number is required. However, other languages, such as C, allow the use of 'macros' to define a string substitution. This is a powerful technique to insert immediate data (no memory reference!) into code while descriptive, symbolic code is used in the source code. Such source code is considered more 'readable.'

To compare the use of immediate data to variable data, Demo *Data03* uses a simple loop with either constant or variable loop boundaries. This is an important application of immediate data, and will be discussed in Chapter 7. Briefly, a loop can have variable boundaries, such as

```
for(x=start_value; x<=stop_value; x+=step_value){
     // working code
}
```

where `start_value`, `stop_value` and `step_value` are variables. Such loops allow functions to be reusable. However, there is a performance cost. The loop boundaries and step must be retrieved from memory. Further, the x86 has a relatively small number of registers (as compared to chips used in traditionally computation oriented mainframes), so the loop boundaries and step may have to be retrieved from memory multiple times. The alternative is to hardcode the loop parameters, which would then require a different loop (or function) for each loop size conceivably needed by the user. This is a very clear performance trade-off between development time and execution time, but there is one added benefit. Explicit, hard-coded loops can be unrolled (see Chapter 7). Raw execution times for Demo *Data03* are shown in Figure 5.3.

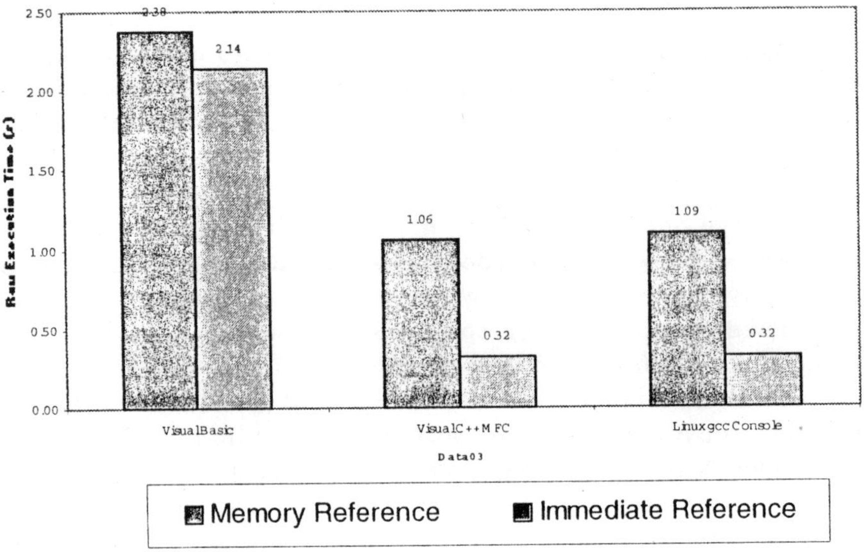

Figure 5.3: Demo *Data03* raw execution times on the test computer for 10,000,000 iterations. Loops that use variable data from memory are compared to those that use immediate, compiled-in, data.

5.4 TYPE SPECIFIC OPERATORS

Some operations in some compilers have 'type specific' operators. An example is the division operator in Visual Basic. Visual Basic includes the 'classic' division operator, /, for dividing floating point data; a separate operator, \, is used for integer division. Other languages, such as C, 'overload' the operators, meaning the same character in source code compiles to different code depending on the context (in C++, function names can be overloaded, too). However, if type specific operators do exist for a particular compiler, it is important to use the proper one; the compiler will insert op-codes for the specific operator appearing in the source code, whether that code is the most efficient or not. For example, in Chapter 2, it was shown that the use of the Floating Point Unit for integer arithmetic is less efficient than using the integer CPU.

To illustrate this, Visual Basic will be used. Demo *Data04* computes int(100/3) three different ways. The first has the two operands declared as floating point values and uses the floating-point division operator, /; the floating point result is truncated to an integer. The second method uses the floating-point operator acting on integer data; the result is truncated to an integer. The third method uses the integer operator on integer data; since this is an integer ALU operation, truncation is implied. Raw execution times for *Data04* are shown in Figure 5.4.

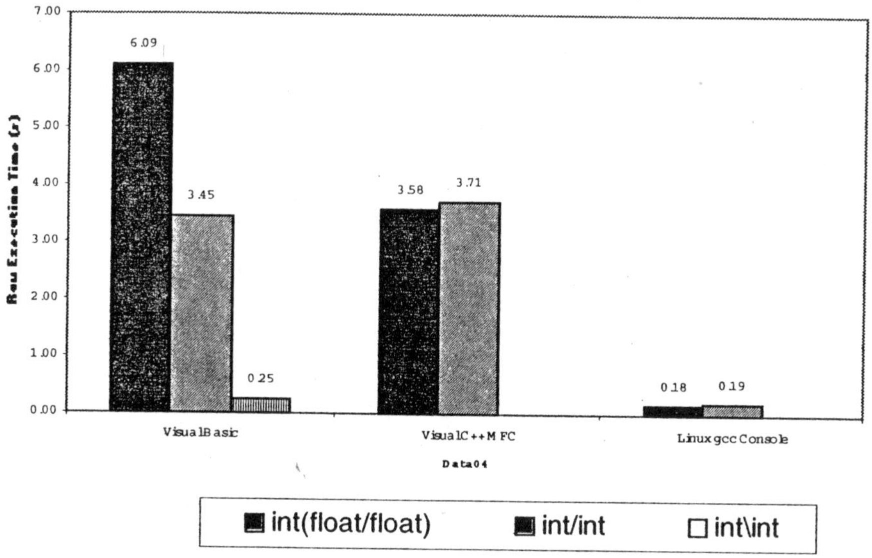

Figure 5.4: Demo *Data04* raw execution times on the test computer for 100,000,000 iterations. Two (C/C++) or Three (Basic) seemingly equivalent division operations that result in a Long Integer are compared.

5.5 VARIABLE SCOPE

It is generally well accepted in modern programming that by declaring variables as *local* as possible creates relatively robust, bug-free, code. That is, in large applications and systems programming, using variables that exist only within the function that used it is considered 'bulletproof' programming. This way, multiple functions (possibly written by different programmers) using the same variable name will not get unpredictable data. However, there are cases where global scope is faster. Demo *Data05* compares the use of local to global data for a simple routine. Raw execution times are shown in Figure 5.5.

5.6 DATA ORGANIZATION

The way that data is organized in memory can influence the performance of programs. Obviously, the key is to organize the data as efficiently as possible for the particular task, given the potential optimization techniques available for each method. As with other aspects of optimization, data organization will often lie on a trade-off continuum between development time and execution performance.

5.6.1 Scalars, Arrays and Associative Arrays

Each memory reference can be given a symbolically unique identifier. Such variables

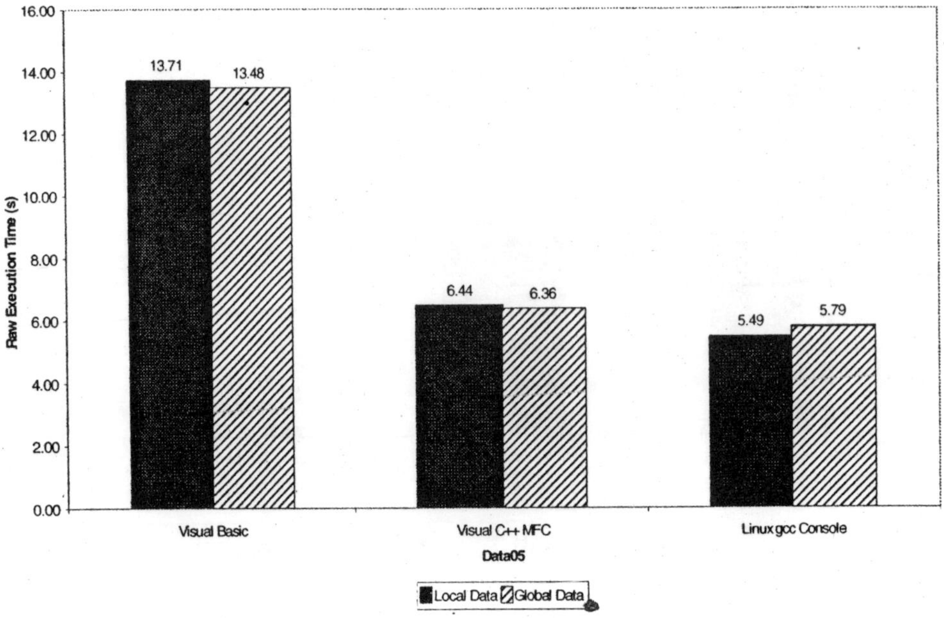

Figure 5.5: Demo *Data05* raw execution times on the test computer for 50,000,000 iterations of the computation of the sine of a random angle. Using the loop counter, angle and sin(angle) result as local data is compared to using global data for these parameters. Note that the performance differential in this demo appears small due to the latency of the random number generator.

would be called scalar. Groups of data given the same symbolic identifier, with members referenced by an index (which compiles to an offset from a base address) are arrays. Arrays are quite amenable to vector, matrix and tensor operations. Another use of arrays is to store functions. The disadvantage of arrays is the data is 'flat;' there are no associations between elements and, say, other arrays. The storage of one-dimensional arrays in memory follow the 'flat' structure.

Multidimensional arrays can be stored in memory several ways, according to the 'order' in which the dimensional indexes are mapped to the linear memory. For example, with C/C++, arrays are stored in 'Row Major' order, as shown in Figure 5.6. This has significant consequences when accessing the data using pointers (such as in ASSEMBLY language) or with the use of Single Instruction Multiple Data instructions. For example, the `MOVQ` MMX instruction moves 8 bytes of data from memory into an MMX register. For the data in Figure 5.6, this could mean moving $a_{00}|a_{01}$ with a single `MOVQ` instruction, but $a_{00}|a_{10}$ could not be moved with a single instruction (these data do not occupy sequential four byte blocks of memory). This is explored again in Chapter 12.

Not all languages store multi-dimensional arrays in a row major or column major linear memory space. Java, for example, stores two-dimensional arrays as *separate* single dimension arrays. The different arrays may not occupy contiguous, or even

Memory Address	Data Value
1000 – 1003	a(0,0)
1004 – 1007	a(0,1)
1008 – 1011	a(1,0)
1012 – 1015	a(1,1)

Figure 5.6: Memory map of a simple 2x2 array stored in Row Major order. The array, a, is single precision, so each element is four bytes.

'nearby' areas of memory. Clearly, this type of language specification increases the pointer overhead required to access random array elements. This is not an approach conducive to high run-time performance involving multi-dimensional arrays.

An associative array allows the 'index' of one array to be a reference to another array. In other words, the *key* of an associative array (the key plays the part of index in a regular, flat array) can be symbolic data. A simple example is to consider the months of the year. Suppose, for a particular scientific application, the average number of hours of sunlight per month is needed. For an array, the initialization may appear as

```
Sunlight_Hours[ 1] = value_for_January;
Sunlight_Hours[ 2] = value_for_February;
Sunlight_Hours[ 3] = value_for_March;
Sunlight_Hours[ 4] = value_for_April;
Sunlight_Hours[ 5] = value_for_May;
Sunlight_Hours[ 6] = value_for_June;
Sunlight_Hours[ 7] = value_for_July;
Sunlight_Hours[ 8] = value_for_August;
Sunlight_Hours[ 9] = value_for_September;
Sunlight_Hours[10] = value_for_October;
Sunlight_Hours[11] = value_for_November;
Sunlight_Hours[12] = value_for_December;
```

To use this array, the programmer must remember that index 1 maps to January, 5 maps to May, etc. An associative array eliminates this by allowing

```
Sunlight_Hours[January]      = value_for_January;
Sunlight_Hours[February]     = value_for_February;
```

```
Sunlight_Hours[March]          = value_for_March;
Sunlight_Hours[April]          = value_for_April;
Sunlight_Hours[May]            = value_for_May;
Sunlight_Hours[June]           = value_for_June;
Sunlight_Hours[July]           = value_for_July;
Sunlight_Hours[August]         = value_for_August;
Sunlight_Hours[September]      = value_for_September;
Sunlight_Hours[October]        = value_for_October;
Sunlight_Hours[November]       = value_for_November;
Sunlight_Hours[December]       = value_for_December;
```

Associative arrays are very useful for systems programming, especially text (such as command line) processing. This can be accomplished in C, for example, by the use of macro's.

5.6.2 Pointers

Arrays are referenced using the index, which at the op-code level generates an offset to the array starting address. Consider a 20 element array of byte data stored at memory locations *arraystart* through *arraystart + 19*. The index of the array then maps to the offset from *arraystart*. This suggests that pointers play an important part in processing array data, though in this case, the pointers may be maintained by the compiler.

There are cases in which the programmer may wish to work directly with pointers, such as using pointer math rather than 'complex' computation of an offset address (as in pixel mapping). As another example, when an array must be passed to (or returned from) a function, the entire array is not pushed onto the stack; a pointer to the array is passed to the function. In C, pointer operators exist (& to generate a pointer and * to reference the data pointed to). Passing an array to a function uses the array name as the pointer. Using pointers to reference array data is generally faster than simple array indexes, such as *Value[I]*.

Pointers allow coding some things that otherwise are very inefficient. For example, the parallel processing techniques discussed in Sections 12.4.2-12.4.3 rely on passing data between compute nodes in the parallel machine. Presumably, this data is single or double precision floating point numeric data, and arrays of numeric data may not be sent on the network using sockets; the data must then be sent as character strings. A point of clarification is warranted: numeric data *can* be sent to the socket, but an array would have to be sent element by element. In network communications, this would imply a separate datagram for each element of the array, meaning each element would have headers transmitted and the flow control between communicating systems would be more complicated. In other words, network bandwidth would be consumed essentially transferring headers and flow control information between the nodes.

Strings of data can be sent as single datagrams, so the bandwidth is utilized for transmission of actual data.

Therefore, the preceding point implies that a means to convert from floating point data to string 'characters' is preferred for the transfer of array data over a network. The C standard library provides several functions useful for generating character strings from numeric data. For example,

```
char *_gcvt( double value, int digits, char *buffer );
```

where value is variable containing numeric data (for purposes of example, 106.442345), digits is the number of digits to write to the string, and buffer is where the string should go. After the function call, buffer therefore contains "106.442345." This string is ten bytes long. A double precision floating point value is stored in eight bytes, so the network bandwidth consumption increases using this numeric-to-string conversion. There is the further disadvantage that this inefficiency grows for higher precision values (such as 106.44234567342, which is still double precision and therefore uses only eight bytes in numeric format).

Section 2.2 outlined the binary representation of floating point data in memory. By reading the four bytes of the single precision data directly, and transferring *those* as strings (or more precisely, as byte arrays), single precision data will always be transferred on the network as four bytes. True, this means a number such as 10 would require four rather than two bytes, but overall, a savings would result. The way to accomplish this 'numeric-to-string conversion' is to generate a pointer to the numeric data, then read four bytes into a byte array.

The Windows API contains a function, CopyMemory (which is actually an alias of RtlMoveMemory, but CopyMemory is more 'readable'),

```
VOID CopyMemory(
   PVOID Destination,      // pointer to destination
   CONST VOID *Source,     // pointer to source
   DWORD Length            // size, in bytes, to copy
);
```

In this function, Destination is the address to which the data is copied, Source is from where it is copied, and Length is the number of bytes to copy. Using this function to perform the numeric-to-string conversion outlined above is illustrated in **Listing 5.7**.

Listing 5.7 Function to convert single precision numeric data to a four byte string

```
void Float2ByteArray(float qWord, char *Abyte)
{
    CopyMemory (&Abyte[0], &qWord, 4);
}
```

This function returns a four-byte array (in Abyte) with each element containing a corresponding byte of the 4-byte data 'qWord.' The byte data in the array can then be used to form a single string via the concatenation operator, '.='. In this fashion, for example, an array of 1000 single precision values can be transferred with a fixed length string of 4000 bytes; this is a reasonable size for a TCO/IP datagram.

Demo *Data06* demonstrates the use of this conversion. For this simple demo, the user types a number into a text box, which upon clicking the command button is converted to a single precision floating-point value. The numeric data is then sent to the function in **Listing 5.7**, which returns the four-element byte array. This array is then displayed as a string representing the number in hexadecimal radix. The conversion is *not* done via a conversion algorithm (such as via Equations 2.2, 2.3 and 2.4), but is simply reading the data stored in floating point format. The reader is encouraged to enter several numeric values (greater than one, less than one, etc).

Once the node transfers the string data, the other node receives the data as a byte array. This introduces another advantage of this method. The receiving node always receives four characters per numeric value (as opposed to using the _gcvt() function, for which the data length may vary). The receiving node contains a function similar to **Listing 5.7** using CopyMemory to convert from the byte array back to the four-byte numeric data. This eliminates the need to either use strtod() or to parse the string to form the numeric data, which would be quite cumbersome compared to the simple conversions using CopyMemory.

To effectively use the CopyMemory or similar function in the manner just outlined, one must know whether the system on which the code will run is *big endian* or *little endian*. This refers to the order in which the four bytes (for single precision) are stored, which is determined by hardware, not OS. Big endian systems (mostly mainframes) store the Most Significant Byte (MSB) in the lower address; PC's are little endian and therefore store the MSB in the higher address. Some systems are actually *bi endian*, meaning they can manipulate data in either endian format.

Another important class of pointers is *function pointers.* These pointers, as one might guess, point to function addresses rather than data addresses. In general programming, functions can be called by name so that address pointers are not necessary. However, C++ allows for an abstraction of function pointers called *virtual function pointers.* These pointers are very important, since they allow a program to determine, at run time, which function to call; this determination need only be made once (rather than iteratively). An example of the need to do this includes graphics routines that contain hardware dependent code; the video adapter present will determine the function that is called. Another example is the CPU dependent extended instructions, as will be demonstrated in Chapter 12. Considering only x86 based architectures, one routine may need to be called if the code is running on an Intel P-III or later, a separate routine if AMD K6 or later and yet a different routine if the code is to use no extended instruction set op-codes. Such an approach eliminates the need for large `switch` or `is` blocks in the working part of the code

5.6.3 Queues and Stacks

Temporary storage of values can easily be accomplished with *queues* and *stacks*. A queue is a First-In, First-Out (FIFO) structure in analogy to a line at a service counter. In contrast, a stack is a Last-In, First-Out (LIFO) structure. Reverse Notation calculators use a stack to 'push' data that is 'popped' when needed. Typically, programming of numerical routines does not require the programmer to create and use these structures.

5.6.4 Linked Lists and Trees

The array is a 'flat' form of data storage that is convenient when the data is to be accessed sequentially or when the index of a random element is known. This method of data storage is extremely inefficient when large volumes of data must be searched for a random element (the index of the desired element is not known). In the linked list, each record in the data contains information about the previous and next records. One extremely important class of linked list is the binary tree. A binary tree has a general schematic structure as shown in Figures 5.7(b) and (c). To define the tree, a definition key is chosen to determine how new data is inserted. The power of the binary tree is the speed with which the data can be searched. A symmetrical tree containing over 1 million records can be searched with as few as 20 comparisons.

The order in which the data is added to a tree impacts its ultimate performance capability. As mentioned, a symmetrical tree is very efficient. However, if the tree is 'linear,' due to an inefficient order of addition, the tree reduces to a linked list, and the searching efficiency is drastically reduced. Therefore, if one intends to use a tree structure to store data that must be rapidly searched, data entry into the tree should be carefully planned. "Apex" data in the tree are termed *roots,* and the binary tree is 'recursive' (each 'branch' also forms a tree 'below' a 'root').

As an example application of a binary tree, consider a sequence of ten bases in a DNA chain (but this data could be an abstract representation of any data that has combinations in 'base-4' radix). These four bases are symbolized by the initials A,T,G

```
AGTGCAAGCT
0123456789
```

(a)

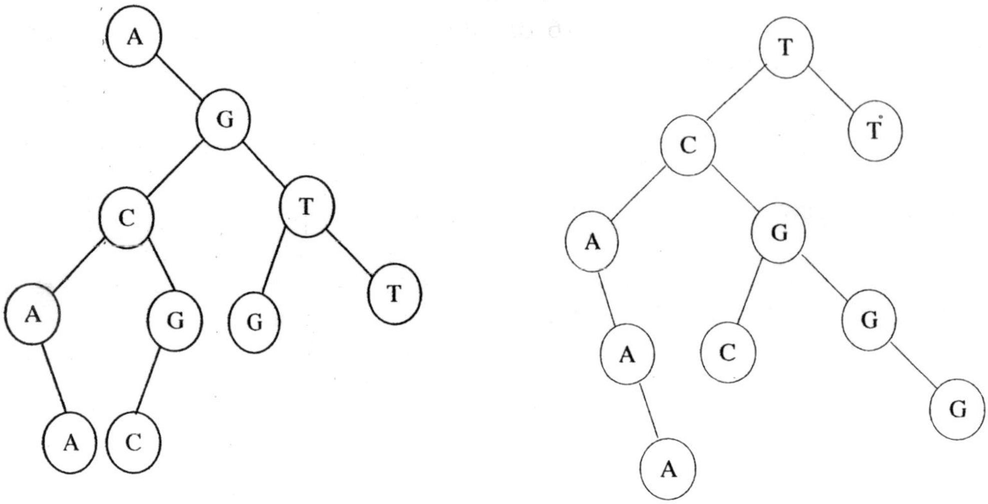

(b) (c)

Figure 5.7: Three methods of storing the ordered sequence "AGTGCAAGCT." In Figure 5.7(a), the data is stored linearly in order, and position numbers are shown (which must be preserved). Figure 5.7(b) and (c) show the data arranged into two possible binary trees. The key for defining the trees is described in the text. The tree in Figure 5.7(b) was constructed by 'forward' application of the definition key to the linear data. 'Backward' application of the definition key was used to construct the tree in Figure 5.7(c).

and C. The sample linear sequence under consideration is shown in Figure 5.7(a), along with 'position numbers' of each base in the sequence. Though numerous binary trees could be defined for this data, two are shown in Figures 5.7(b) and 5.7(c), respectively. To construct a tree, a value is compared to a root, then moves or inserts right or left depending on the definition key. The same definition key was used for both trees in Figure 5.7:

> If data < 'root', move or insert LEFT
> If data >= 'root', move or insert RIGHT

The tree in Figure 5.7(b) was constructed using this key applied forward (positions 0-9). Backward (positions 9-0) application of the definition key was used to construct

Table 5.1: Number of comparisons required for finding data that meets criteria given in "Search Key" (chosen at random) for several searches of the data in Figure 5.7. The tree searches generally require fewer comparisons to find the desired data.

Data Storage	Search Key and Number of Comparisons First "C"	 Second "A"	 Second "T"
Linear, Forward Search, Fig. 5.7(a)	5	6	10
Linear, Backward Search, Fig. 5.7(a)	2	5	8
Tree, Fig. 5.7(b)	3	4	4
Tree, Fig. 5.7(c)	2	4	2

Figure 5.7(c). It should be noted how different the trees are for the different *order* the data is presented, *even though the same definition key is used.*

Searching this data for specific occurrences of particular bases serves to show the relative efficiency of the tree structure. Several search keys and the resulting number of comparisons required to find the key are shown in Table 5.1. Notice that in the general case the tree requires as few as $1/5^{th}$ the number of comparisons, even for this simple example. For longer strings of data, and more complicated search keys, the relative efficiency of the tree may be even greater.

5.6.5 Structures-User Defined Data Types

One very useful technique of data abstraction is the use of *structures*. Structures allow the grouping of data references so that the entire set can be referenced with a single symbol. This powerful technique is useful to pass a large amount of data to a function by actually passing only a single pointer (as discussed in Chapter 6). As a simple example, consider an ATOM structure as might be used in an electrostatic modeling calculation:

```
typedef struct ATOM_TYP
{
      int   NuclearCharge;
      float X;
```

```
    float  Y;
    float  Z;

} ATOM,  *ATOM_PTR;
```

Structures can contain arrays, there can be arrays of structures, and they can have other structures as elements. As an example of a nested structure, consider the MOLECULE structure that contains ATOM defined above:

```
typedef  struct  MOLECULE_TYP
{
    int         NumberOfAtoms;
    int         Charge;
    float       CenterX,  CenterY,  CenterZ;
    ATOM        *atomlist;

} MOLECULE,  *MOLECULE_PTR
```

Notice that the ATOM type was used as a data type (that gives the positions and nuclear charges of the individual atoms). The variable NumberOfAtoms specifies the number of atoms, Charge is the net molecular charge and CenterX, CenterY and CenterZ are the x, y and z coordinates of the center of the molecule.

To use the nested structure to represent, say, a water molecule, the variable Water is declared as type MOLECULE. Since the MOLECULE type contains a dynamic data (the number of atoms is not specified in the type definition), the malloc() function should be used following the declaration of Water (in C++, the new() function can be used). An example for the Water molecule, H_2O, is shown in **Listing 5.8**.

Listing 5.8 Declaration and Initialization of Water Variable using Type MOLECULE

```
MOLECULE Water;

//initialize the neutral Water
Water.NumberOfAtoms= 3;
Water.Charge       = 0;

//initialize the pointer to the atom list
Water.atomlist
=(ATOM_PTR)malloc(Water.NumberOfAtoms*sizeof(ATOM));
```

```
//assign charges and positions of atoms
//for a guess geometry
//all positions in angstroms
//charges in atomic units (not coulombs)
//oxygen
Water.atomlist[0].NuclearCharge   = 8
Water.atomlist[0].X               = 0;
Water.atomlist[0].Y               = 0;
Water.atomlist[0].Z               = 0;

//hydrogen 1
Water.atomlist[1].NuclearCharge   = 1;
Water.atomlist[1].X               = 0.79;
Water.atomlist[1].Y               = 0.62;
Water.atomlist[1].Z               = 0;

//hydrogen 2
Water.atomlist[2].NuclearCharge   = 1;
Water.atomlist[2].X               = -0.79;
Water.atomlist[2].Y               = 0.62;
Water.atomlist[2].Z               = 0;

//for simplicity, specify the oxygen atom to be the
//'center'; one could use center of mass, etc.
Water.CenterX      = Water.atomlist[0].X;
Water.CenterY      = Water.atomlist[0].Y;
Water.CenterZ      = Water.atomlist[0].Z;
```

This single data reference, Water, contains information regarding the number of atoms, their positions and their charges as well as the location of the molecule. Individual member references are made by the "dot" notation. (If one generates a pointer to the structure, individual member references can be made with the arrow operator, ->).

While this method of data organization is very efficient for the programmer, there can be a performance cost. Accessing the oxygen atom charge as Water.atomlist[0].NuclearCharge is relatively slow. If this value is needed repeatedly in code, especially in a large iterative loop, faster code results if the value is *cached* into a simple scalar variable. For example, in Demo *Data07*, a simple structure is

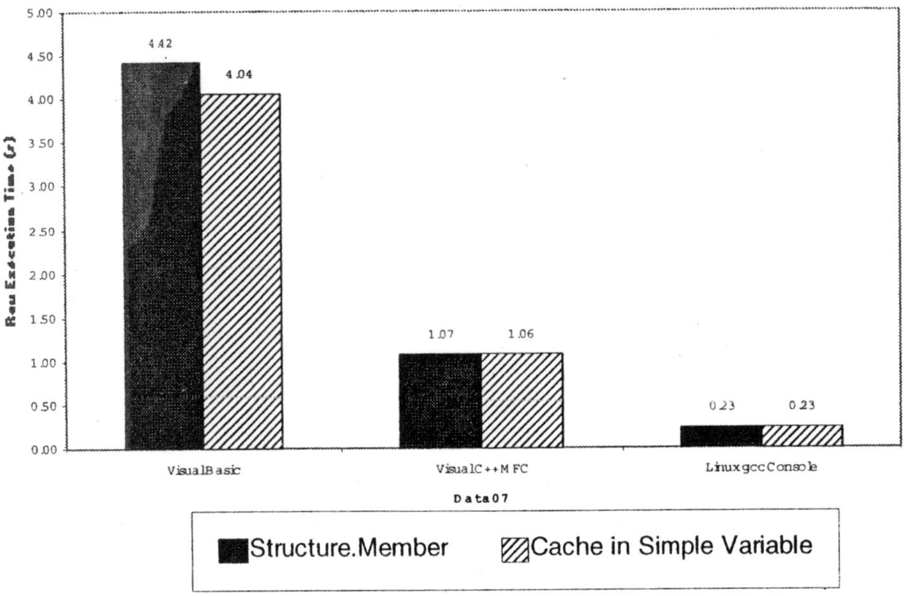

Figure 5.8: Demo *Data07* raw execution times for 100,000,000 iterations on the test computer. In this demo, using structure (or object) member access is compared to caching the data into simple variables.

created and some basic floating-point math is performed. Raw execution times for *Data07* are shown in Figure 5.8. Though this gain is small, this is very simplistic sample code; complicated structures in complicated mathematical routines may show a considerable efficiency difference between member access and caching.

5.6.6 Objects and Classes

Object oriented programming is a very powerful technique of data abstraction. Formally, objects are data units that *encapsulate* data and relevant code so that manipulations are *part* of the object. Consider as an example a command button Windows GUI object. When a programmer uses a command button on a form, the command button object *contains* the code to render the button, send messages to the OS (such as when the button is clicked or the mouse cursor is hovered over it), to enable or disable it, to print text on it, etc. The programmer does not need to write this code every time a command button is needed, but only need provide data and event handlers to process the events created by the object. While this book is not a tutorial for object-oriented programming techniques (there are numerous books on this subject), a brief discussion is warranted as relevant to performance issues.

The data abstraction that comes with object-oriented code is a powerful means of code reuse, application integration and rapid development. By containing their own code, objects are, in a sense, self-contained units. However, the system overhead associated with object-oriented programming can be large. Therefore, when execution speed is desired, the effects of objects should be considered. Indeed, in Tricks of the Windows Game Programming Gurus, Andre LaMothe recommends against the use of classes (and therefore objects) in high performance game programs unless the programmer is expert in their implementation (except those objects and classes necessary to interact with Windows, of course).

However, when the proficiency is present, some key optimizations can occur when programming with objects. Todd Veldhuizen, in a paper outlined at http://osl.iu.edu/~tveldhui/papers/DrDobbs2/drdobbs2.html, presents the definition of several classes (and operator overloading) that demonstrates C++'s superior performance to Fortran, especially when scalability is an issue. Veldhuizen argues that key in these optimizations are compiler advances in C++, so it remains arguable that *older* Fortran generally outperformed *older* C code. Admittedly, however, he was using a highly optimized C++ compiler and comparing to Fortran 77. The point is that high performance was achieved by using object-oriented programming.

The basis of an object is its class. A class defines the data and code of an object and it is often said that an object is an instance of a class. Specifically, a class defines the properties (data), functions (methods) and function calls (events) of an object.

Further, classes can inherit other classes, so that classes that are more specific do not need to 'rewrite' common parts of the class. For example, one may have a PERSON class whose code and data are inherited by the MAN class, the WOMAN class and the CHILD class.

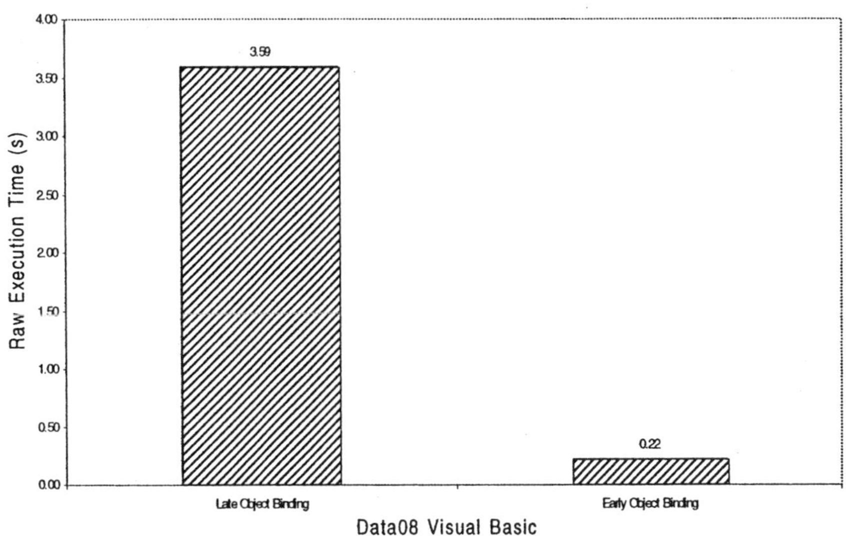

Figure 5.9: Demo *Data08* raw execution times for 1,000,000 iterations on the test computer. This demo compares the performance for late object binding to early object binding in Visual Basic.

Objects can be late bound or early bound. This refers to when the object is actually 'created.' Early bound objects are created at compile time, so their code exists within the executable. As such, code using early bound objects executes faster. Late bound objects are created at run-time, and generally do not execute as fast as early bound objects. A comparison of the run-time performance of early to late bound objects is shown for Visual Basic in Figure 5.9. Whether early or late bound, there is considerable operating system overhead when working with objects since memory management issues are far more complex than when using simple variables. In either case, it should be noted that, as for structures, object data references should be cached into simple variables when performance is needed. In C++, late binding is achieved by the use of virtual functions, but this is not demonstrated in this book.

An example of using an object in a scientific application extends the example from the previous section. Suppose the MOLECULE is defined as a class with similar member data as the structure above. In the sample code above, the center of the molecule was chosen to be the oxygen atom; however, a better choice for 'center' might be the center of mass. This can be computed in a member function that is part of the class itself. In other words, the code to compute the center of mass is encapsulated in the Water object along with the data storing positions and charges. Doing this means that any programmer using the MOLECULE class automatically has the function to compute the center of mass; no explicit function call is needed, for example.

5.7 DATA ALIGNMENT

The address at which a piece of data is located in memory can greatly influence the speed with which the data can be accessed. With data types intrinsic to the language, such as integers, floats and doubles, the data should lie on natural addresses to the type. For example, word data should lie at even addresses, double word data should lie at addresses divisible by four, etc. This is simple enough for intrinsic types. However, alignment becomes an important issue with structures and objects.

Consider a structure consisting of two long integers, a short integer and two floats. If the structure is organized in this manner (long, long, short, float, float), the data is generally *not* aligned. Consider an array of these structures, as shown in Figure 5.10(a). Clearly, the double word data in the second element is not aligned. This part of the alignment problem can be eliminated by padding the structure (perhaps with a short), so that the first long integer is always aligned. However, even in this case, the float data will not lie at addresses divisible by four, so there remains an alignment issue. This structure would better be organized as: long, long, float, float, short, short (for padding). This is shown in Figure 5.10(b).

Alignment is of such importance that some performance related instructions *require* aligned data. For example, the Streaming SIMD Extensions (SSE) instruction set on

	Address	Data Type
	1000	Long
	1004	Long
	1008	Short
Not Divisible by 4	1010	Float
Not Divisible by 4	1014	Float
Not Divisible by 4	1018	Long
Not Divisible by 4	1022	Long
	1026	Short
	1028	Float
	1032	Float

(a)

All double words at addresses divisible by 4

Address	Data Type
1000	Long
1004	Long
1008	Float
1012	Float
1016	Short
1018	Short
1020	Long
1024	Long
1028	Float
1032	Float
1036	Short
1036	Short

(b)

Figure 5.10 (a) Two structures consisting of the intrinsic types indicated stored in memory with no planning for alignment of the data. (b) The structures redesigned to be properly aligned with respect to the member data.

the Intel Pentium III (and newer) CPU's have an instruction for moving packed aligned data (`MOVAPS`) and packed unaligned data (`MOVUPS`). The former is *always* faster, but cannot be used for unaligned data. Many compilers have switches or preprocessor directives to help automate the alignment of data, but the programmer should organize structures and objects in a way conducive to proper alignment.

5.8 NAMING VARIABLES, FUNCTIONS, OBJECTS AND CLASSES

Key to code organization is having a systematic naming convention for data and functions. Anyone familiar with Microsoft documentation has probably observed what is, at first glance, a cryptic language. Variables are named to give an indication of type and/or scope; classes begin with "C," etc. It should be pointed out that intrinsic to this Hungarian Notation, and its variants, is the premise that multiple programmers may be working on a single project. Therefore, it is key that two programmers can easily interpret the name for a given piece of data.

This author does not program in the Microsoft Hungarian Notation, but does follow some similar conventions. This is not to suggest it is inefficient, but rather that some notations in the scientific or mathematical context actually may increase confusion. Consider, for example, code that references a function stored as an array, *R*, and its total first derivative in another array, *dR* . In the Hungarian notation, if R is double precision, it 'should be' named as dR. This could easily be confused (or mistyped) with the derivative, which may be something like ddR ("d" for double precision, dR as variable name). Further, Microsoft does not suggest the use of underscore, _, in function names; the author finds underscores to be very useful delimiters, but this is just personal preference.

The variable naming convention used throughout this text (and generally used by this author) for mathematical routines is rather simple. Variables are generally named as they appear in equations: *x, y, r, E*, etc. Sometimes, the full name is used, such as *Energy*. Reciprocal values are indicated by underscore-dividend. For example, *1/r* is represented at r_1. An exponent is represented by simply concatenating the exponent (if a simple integer) to the variable name; r^6 is named as r6; the same notation is used for subscripts provided no name 'collision' occurs. Multiplicative factors are named by x-factor, such Distancex4 for 4 * *Distance*. This method is easy to remember and symbolically captures the meaning of the variable. It is nowhere near as robust as the Hungarian notation, but serves a more specific purpose.

Whether a programmer follows Microsoft's, or any other, specific notation is immaterial. What does matter is that the programmer and anyone reading the source code (including Copyright Office staff) can interpret what the names represent. One technique to minimize potential confusion is to include clear, informative comments in the code.

Chapter 6: Function and Procedure Calling: Optimizing Program Flow

Many computer programs consist of repetitive tasks, and this certainly applies to scientific applications. These tasks, divided into logical units, are often executed in sequence, but may sometimes be event driven, or even called as needed by the control portion of the software. Though such logical units are useful for code modularity and reuse, the misuse of these units will dampen a program's speed of execution.

These reusable units are traditionally either functions or procedures, though modern programming styles also achieve code reuse via data object and component object models. Once collectively called subroutines, functions and procedures differ mainly by their *exit state*: functions return a value 'in their name,' procedures do not. For this reason, functions provide the greater power since exit states can be tested and used for error checking, but there is additional overhead.

In this Chapter, the mechanisms involved in subroutine calling will be examined, followed by examples of an inefficient programming style. Optimizations and programmatic improvements related to function calling (in the interest of conciseness, function and procedure calling will be referred to as simply function calling) will be treated explicitly, though compiler switches may be available to let the compiler do the optimization. Issues related to function library building, hand coding intrinsic or standard library functions and a comparison of iterative vs. recursive function calls will be presented. Numerous numerical computation library resources are listed in Appendix A.

6.1 MECHANISM OF FUNCTION CALLING AND INLINE CODE

As discussed in Chapter 2, program flow is ultimately maintained by the Instruction Pointer register in the processor. During a given clock cycle, the Instruction Pointer contains the address of the instruction being executed, and maintaining the Instruction Pointer is the job of the processor itself; there are no instructions in the x86 instruction set to modify its value programmatically. Any branches in program flow, as created by ASSEMBLY `JMP` instructions for example, require the processor to change the Instruction Pointer appropriately.

In a simple, non-protected memory system, at least two 'jump' types can occur: near and far. The memory addressing scheme in the segmented, non-protected 8088 based system, as discussed in Chapter 2, utilizes two values to define an address. Since the Segment Offset is 16 bits, each segment contains 64 kB; any jumps within a segment require only a change of the offset stored in the Instruction Pointer. However, if the program flow requires a jump beyond 64 kB, both the Instruction Pointer *and* the Code Segment register must be updated. Therefore, a Far jump will require additional work for the processor than will a Near jump. The flat 32 bit addressing used on modern systems eliminates this complexity somewhat. Part of the advantage of the newer flat 32 bit addressing mode can be seen; for any address function, 32 bits must be manipulated but the mechanism is always the same.

A function call differs from a simple branch several ways. First is that the return address must be stored. In this sense, the function call may be thought as a "Jump" and "Jump Back." This introduces the key issue with function calling: stack manipulations. A function call must also 'create' a stack space (generally within the stack space of the calling program) for local data, and any parameters passed to or from the function must be placed on the calling program's stack. At the very least, a function call in a high level language must PUSH the return address onto the stack (or store it *somehow*) and POP the address when the function is finished. Consider the code in **Listing 6.1**.

Listing 6.1 **Example of Code with function call**

```
// previous code

A = 1.234;
B = X2(A);

// next line of code
// a bunch of code

float X2(Value as Single){

      float Factor;
      Factor = 2;

      return(Value * Factor);

}
```

Here, code execution branches at the B = X2(A); statement (which is a call to the function X2), but the address of the NEXT LINE OF CODE must be stored somehow. When execution reaches the `return` statement, a `RET` to the address of 'next line of code' can be made; this is usually via an offset from the `RET` op-code within the function. Thus, every simple procedure call, with no parameters passed and no return value, involves at least two execution branches.

Stack manipulations must be done in the basic procedure call: no parameters passed and no exit state returned. With the passing of parameters (either entry or exit), the stack is also used as the scratch pad between programmatic execution units. Further, for a high level language, local variables used in the subroutine are generally allocated out of the program's stack space, rather than general memory. Therefore, when parameters are passed or if the subroutine defines its own set of local variables, the stack manipulations increase. Again, recall that at the hardware level, each stack manipulation requires updating the top-of-stack pointer in addition to transferring the actual data.

Consider a program that repetitively calls a function, for example, inside a large loop. *Each* `CALL` requires storing the return address. If parameters are passed, or if the function declares its own local variables, each `CALL` also requires repeated changing of the stack pointer along with `PUSH`ing and `POP`ing the data. These manipulations necessary for calling the function constitute wasted CPU cycles. The alternative is to use inline functions: the code is explicitly present, in its order of execution, so no `CALL` is made. Inline code will generally make the code larger, which is another example of the trade-off between size and speed optimization.

A simple memory map showing calling and function stacks for the rudimentary function call sequence in **Listing 6.1** is schematically shown in Figure 6.1. In Figure 6.1(a), just before the call is made, memory has been allocated for the variables *A* and

FIGURE 6.1

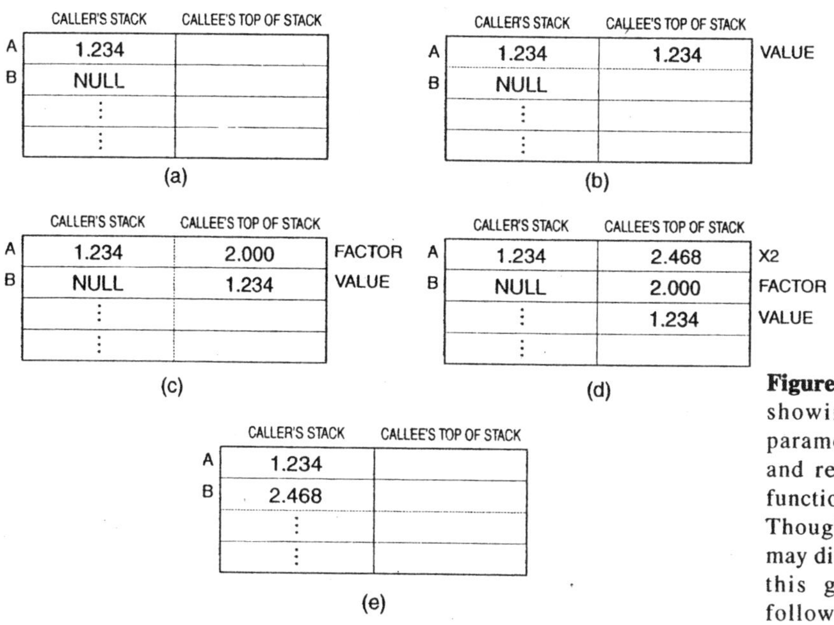

Figure 6.1: Schematic memory map showing the location of passed parameters, locally declared data and return values for the simple function call shown in **Listing 6.1**. Though languages (or compilers) may differ in details, something like this general sequence will be followed.

B and a stack space has been designated for the function X2. When the actual calling line is executed, the passed parameter, *A*=1.234 (labeled VALUE in the function itself), is placed on the function's stack; this is shown in Figure 6.1(b). Figure 6.1(c) shows the stack allocation of the function local variable FACTOR as the function executes and the local variable is declared. As the function further executes, it generates a value for 'itself,' which is the result of the working part of the function; this is shown in Figure 6.1(d). Finally, Figure 6.1(e) shows the stacks after the returning from the function call; the stack for the function is cleared and the desired result is stored in variable *B*. It is worth noting that these are not separate stack spaces; the function's stack is actually on top of the calling program's stack. Therefore, it is important to understand that following any function, the stack must be 'cleaned' of any extraneous values else the calling program's stack pointer will not point to the proper value.

Demo *Function01* demonstrates the overhead associated with function calling. In this demo, a DOUBLE is passed to a function that declares two local variables and returns a DOUBLE. The function does no actual work. The function call is wrapped by a loop that calls the function 100,000,000 times. Execution times for the CALL version and the Inline version are shown in Figure 6.2. Shown in Figure 6.3 are execution times for Demo *Function02* in which some computational work is done. This is to show that the advantages of Inline code are not an artifact of calling an empty routine.

Of particular interest in Demo *Function01* is the declaration of the local variables. When the function is called 100,000,000 times, the local variables must be allocated (and destroyed) on the stack 100,000,000 times. For the Inline version, the 'local variables' are allocated one time, though the computation is performed 100,000,000 times. In any case, by eliminating the combined stack manipulations, the Inline code

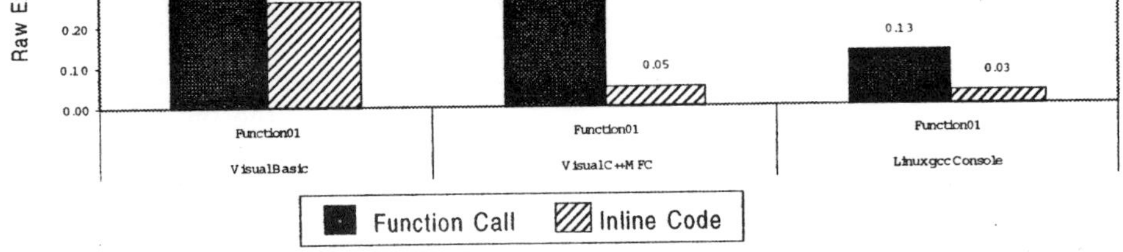

Figure 6.2: Demo *Function01* raw execution times on the test computer for 100,000,000 iterations for a simple (empty) function call compared to inline code. No work is done in either the called or inline functions.

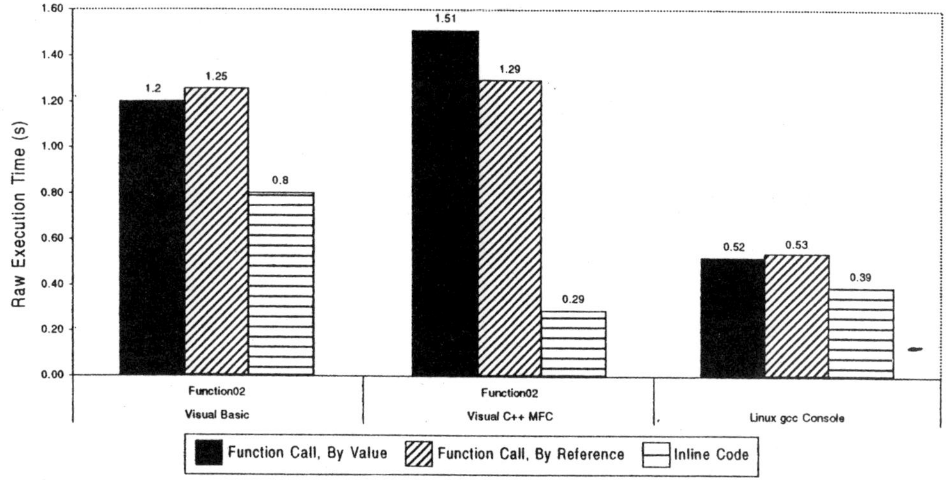

Figure 6.3: Demo *Function02* raw execution times on the test computer for 10,000,000 iterations for a simple function call compared to inline code. A simple Lennard Jones pair energy calculation is done in this case to show that the Inline advantage is not an artifact of the empty functions depicted in Figure 6.2.

is faster by a factor of 3. A three-fold increase may not be considered much. However, only one parameter is passed in this demo.

Demo *Function03* illustrates the cost of parameter passing. In this demo, the code passes two, four and eight doubles and no work is done in the functions. The execution time versus number of parameters is shown in Figure 6.4. Notice the general linear (Number of Parameters) dependence. This shows that the function calling performance degrades as the number of parameters increases.

One might suggest that rarely in computational routines are eight double precision values passed. This may be true for general computational functions, but keep in mind that many Windows API functions require many parameters, some of which may be complex data structures. Further, the author has examined random lines of code in random source files for a common, widely distributed *ab initio* quantum chemistry package. Two separate source files (of the three examined) contained functions called with multiple double precision values passed. One call included eight doubles, one had seven doubles, another had six doubles and one had an array reference and five doubles (three of which were defined as ZERO!). In addition, there are scientific applications where making a large number of values available to a function is required. Such applications may be a linear equation solver, a matrix inverter, eigenvalue solver, etc. In these cases, the data is generally held in arrays.

In the case of a function working on an array, one typically passes a pointer to the array, rather than passing each value directly. Sample code *Function04* compares calling a function by passing six array elements vs. passing the single pointer to the function. Raw execution times for demo *Function04* are shown in Fig. 6.5. Though the performance difference in this simple demo is not very dramatic, for larger arrays

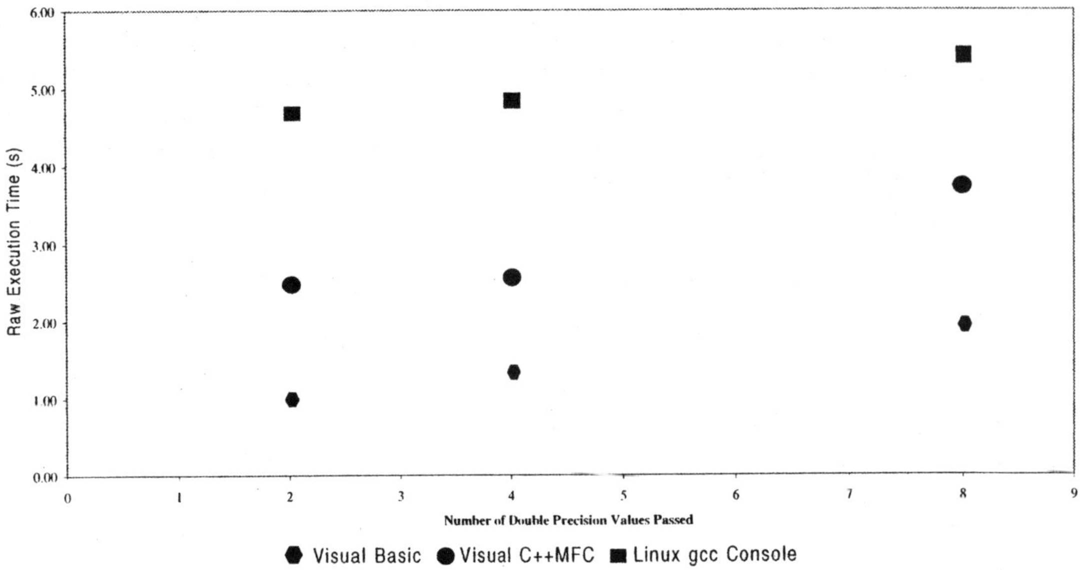

Figure 6.4: Demo *Function03* raw execution times versus number of double precision floating-point parameters passed to simple (empty) functions.

Figure 6.5

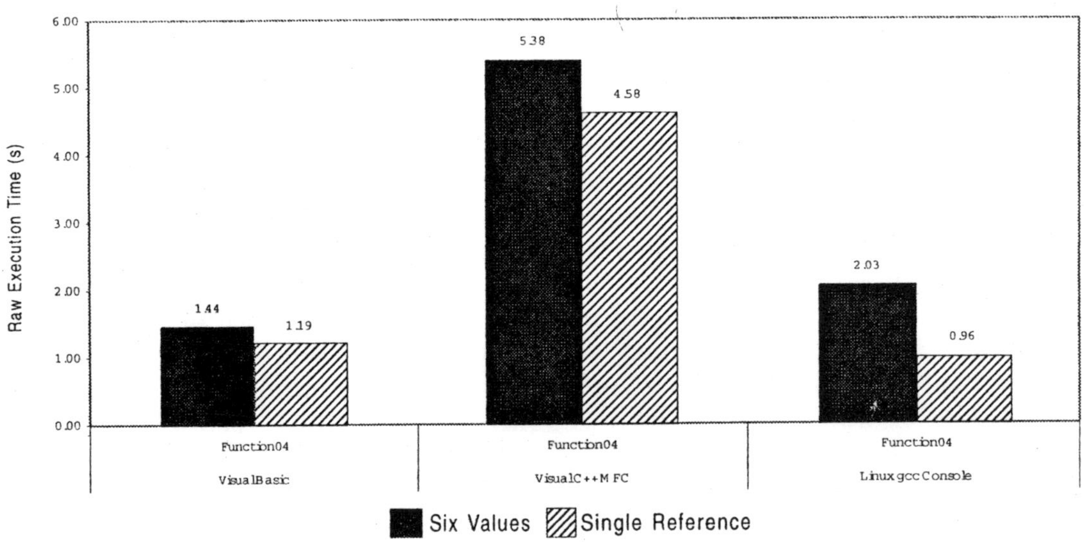

Figure 6.5: Raw execution times (Demo *Function04*) for 100,000,000 iterations for a simple function call. A function call passing six double precision floating-point values is compared to a call passing a single array reference. No work is done in either of the functions.

the difference would certainly be greater. This demo suggests an interesting optimization even when the passed data are not 'grouped' in an array structure. The passed parameters can be stored in an array and a single pointer is passed. In this fashion, a large number of parameters can be passed at the cost of only a single pointer.

Consider for example a 3-dimensional distance function to which the six coordinates *x1*, *x2*, *y1*, *y2*, *z1* and *z2* are passed. Passing each of these values individually, each time the function is called executes approximately 18% slower than if the six values are placed in an array and the reference is passed. There may be cases in which the overhead associated with maintaining this array structure outweighs the advantage; again, each application should be examined since there is no general rule.

6.2 PROGRAMMING IN THE "SUB-ROUTINE STYLE"

To begin this section, a programming style is presented that the author has encountered many times in scientific codes. This style makes heavy use of sub-routines; each function is delegated to a sub-routine, which is called as needed. This style does have several advantages to the programmer: code associated with each functional unit is written only once, smaller source and compiled code and modularization/reuse of code. It should be noted that these advantages correspond to the Scientist Programming Style.

An extreme version of this 'subroutine' style is illustrated for the Monte Carlo procedure in **Listing 6.2**, which follows the flow chart shown in Figure 1.1. Once again, notice that the Scientist Style follows a nearly direct 'translation' from Flow Chart to code. Each of the called functions are type void and thus return no value.

Listing 6.2 Scientist "Subroutine" Style of Programming

```
// Function to assign initial
// particle positions
Initial_Position();

// Function to compute the total
// system energy
Calc_Total_Energy();

For(configuration=1; configuration <= NumberOfConfigurations;
    ++configuration){

    // Function to move a random
    // particle
    Move_Particle();
```

```
      // Function to compute the
      // energy change upon moving
      // the particle
      Calculate_Delta_E();

      // Function to test the new
      // energy against the chosen
      // distribution function
      Test_Move();

      // keep if move okay
      if (MoveOK) {

            // code to update particle
            // positions to keep the new
            // configuration

      }

}

// Function to output result
Print_Energy();
```

Listing 6.2 has the modular advantage that if the various functions are placed in a compiled library, the functions can be modified without having to recompile the main executable. This is a big advantage when dealing with distributed applications. From an overall development standpoint, this may be desirable, but the overhead of calling each function when needed should be considered.

The alternative to **Listing 6.2** for the basic Monte Carlo procedure is shown in **Listing 6.3**. **Listing 6.3** utilizes 'in-line' functions; the code to perform the various tasks actually appears in sequence. All calls are eliminated.

Listing 6.3 **Coding functionality inline rather than called subroutines**

```
// Actual code to compute Initial Position
// Actual code to compute Total Energy

for(configuration=1; configuration <= NumberOfConfigurations;
++configuration){

      // Actual code to Move a Particle

      // Actual code to compute the energy
      // change
```

```
        // Actual code to test the move

        // keep if move okay
        If (MoveOK) {

                // code to update particle
                // positions to keep the new
                // configuration

        }
}

// Actual code to Print Output
```

That **Listing 6.3** is an optimization can be shown explicitly by computing the number of function calls eliminated. In **Listing 6.2**, there are four calls within the ConfigurationCounter loop and three outside the loop. Therefore, **Listing 6.2** contains

Eq. (6.1) $$4 \cdot (\text{NumberOfConfigurations}) + 3$$

function calls. In the Monte Carlo procedure, reasonable statistics do not accumulate for even simple monoatomic isotropic systems with NumberOfConfigurations less than about 30,000. For such a basic calculation, **Listing 6.3** reduces branching overhead by about 120,000 calls!

Each call requires overhead. Returning attention for a moment to Demo *Function01*, the ASSEMBLY Language output from the C compiler is given in **Listing 6.4** for the actual function call.

Listing 6.4 **MS VC++ ASSEMBLY Language output for C function call**

```
//call the function
energy = calc_ljenergy(r);

        mov   eax, DWORD PTR _r$[ebp+4]
        push  eax
        mov   ecx, DWORD PTR _r$[ebp]
        push  ecx
        call  ?calc_ljenergy@@YANN@Z
        add   esp, 8
        fstp  QWORD PTR _energy$[ebp]
```

Each call requires pushing the passed parameter (double precision r in this case) onto the stack, and the return value is returned in the x87 register st0. Specifically, the call requires

```
2     PUSH register          for a total of   2 Pentium cycles
1     CALL far (external)    for a total of   4 Pentium cycles
1     FSTP 64 bit mem        for a total of   1 Pentium cycles
1     ADD  register, imm     for a total of   1 Pentium cycles
2     MOV reg, 32 bit mem    for a total of   2 Pentium cycles
```

which is a total of 10 cycles just to make the function call. In the function itself, and for *every* function called, the return address must be stored. Therefore, code like the ASSEMBLY Language in **Listing 6.5** is produced for every function.

Listing 6.5 MS VC++ ASSEMBLY Language output for called C function

```
;this occurs on entry into the function
;calc_ljenergy(double)
push ebp
mov  ebp, esp
sub  esp, 80              ; 00000050H
push ebx
push esi
push edi
lea  edi, DWORD PTR [ebp-80]
mov  ecx, 20              ; 00000014H
mov  eax, -858993460; ccccccccH
rep stosd

;after the function is done, and
;'returns' its value to the calling
;routine, this occurs before execution
;returns to the caller
pop  edi
pop  esi
pop  ebx
mov  esp, ebp
pop  ebp

;actually returns control to caller
ret  0
```

This sequence constitutes

4	PUSH reg	for a total of	4	Pentium cycles
2	MOV reg, register	for a total of	2	Pentium cycles
1	LEA reg, 32 bit mem	for a total of	1	Pentium cycles
2	MOV reg, imm	for a total of	2	Pentium cycles

1	REP STOSD(20)	for a total of	23	Pentium cycles
4	POP reg	for a total of	4	Pentium cycles
1	SUB reg, immed	for a total of	1	Pentium cycles
1	RET near	for a total of	2	Pentium cycles

Combining this returning code with the calling code in **Listing 6.4**, there is a total of 49 cycles *for each function call.* The Monte Carlo procedure in **Listing 6.2** was to make 120,000 calls (for 30,000 configurations) which results in 5.88 megacycles absolutely wasted on calling overhead in a situation that such overhead gives virtually nothing. This situation is even worse when more parameters are passed and more complicated stack manipulation instructions are generated by the compiler.

One may argue that 5-6 megacycles is not a lot in terms of modern processors running at gigacycles per second. One should remember that these latencies are for op-codes that are *already* fetched and decoded, so represent *best-case* scenario. Further, it should be emphasized that this example was not very large and does not include parameter passing and local variable stack allocations. It is common in computational simulations to have far more than 120,000 function calls. Repetitions of a procedure in a parameter study, a temperature effect study (or any real research project involving an iterative procedure) would add to the wasted cycle overhead. The key point is that the CPU is not accomplishing any useful work by calling and returning from a function.

6.3 CALLING WITH REDUCED STACK OVERHEAD

How does one eliminate passing entry and exit parameters? The easy answer is with global data. A trade-off must be considered, however: using local variables is one way to keep the program 'bullet-proof.' That is, the general subroutine style using local variables has been shown to produce code that has fewer bugs and is easier to debug. Therefore, though the use of globals is often faster, their use should be carefully weighed in the bigger issue of buggy code does not solve problems faster. A general guideline is to use globals *when you can*, and this will be determined by each project. Further, one should avoid the use of globals that are manipulated by many functions, and careful, systematic variable naming is extremely important. Another drawback of globals may be larger overall memory consumption.

For example, consider **Listing 6.6** and **Listing 6.7**, which contrast the parameter passing vs. GLOBAL scope techniques.

Listing 6.6 Calling a function by passing a parameter

```
Calc_Delta_Energy (TheMovedParticle);

double Calc_Delta_Energy (intParticleMoved){
```

```
      // Code to compute and return the
      // Lennard-Jones Energy change

}
```

Listing 6.7 Calling a function that uses GLOBAL data

```
// in global declaration section
int ParticleMoved;
double Delta_E;

// ParticleMoved is set in code

double Calc_Delta_Energy(){

      // code to compute the Lennard-Jones
      // Energy change using the GLOBAL
      // variable ParticleMoved, sets the
      // computed value to the GLOBAL
      // variable Delta_E}

}
```

Execution of a program using **Listing 6.6** requires, in addition to the basic calling overhead (preserving the calling state of the CPU) a `PUSH` of the data contained at the address referenced by the variable name ParticleMoved. Since ParticleMoved is an int, four bytes are `PUSH`ed and four are subsequently `POP`ped. The data is not `PUSH`ed or `POP`ped in **Listing 6.7** since it is modified directly in memory. Demo *Function05* compares passing local data to the use of globals, and the empirical results are summarized in Figure 6.6.

Many languages provide a `static` declaration qualifier. This keyword in the declaration of local data prevents the destruction of that data when a function exits. By declaring local variables as `static` may reduce the stack overhead incurred in function calling, but cannot eliminate it. In this regard, `static` variables should probably be treated like globals and be properly initialized when needed since the use of `static` may create hard-to-find bugs if a function is called from many different calling functions. However, due to scoping rules, `static` may help eliminate bugs if several functions happen to utilize the same (local) variable names. This is an advantage that locally declared static variables have over using global data.

Object oriented programming may reduce the stack overhead involved in parameter passing. When one calls a method, parameters are often not passed; the data the method needs is encapsulated (as public properties and private data) with the object. Rather than passing parameters, one can set properties prior to the method 'call.' The net performance is about the same, however, since one must add extra lines of source to set the properties.

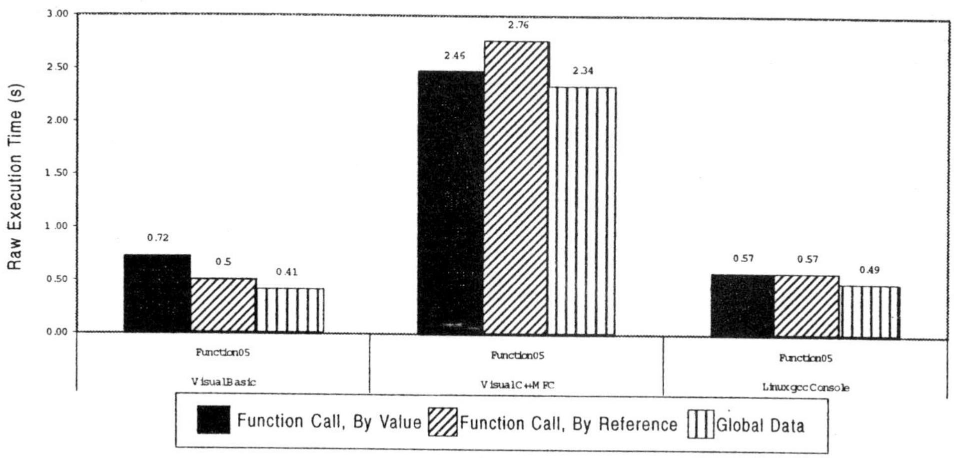

Figure 6.6: Raw execution times (Demo *Function05*) for 50,000,000 iterations for a simple function call. Function calls passing a value or reference is compared to a function call that works on global data. No work is done in either of the functions.

6.4 USING LIBRARY FUNCTIONS

Given the potential for the inline performance improvements summarized in Figures 6.2 and 6.3, one may ask, "Why is so much programming done using functions and procedures?" The answer is simple: modularity. The subroutine style offers several opportunities to reduce development time, but these should be carefully weighed within the performance context of a particular program.

Modularity allows a programmer to write a function or procedure once, and not just for calling in one program. The code may be used by many programs, so long as the prototype is known to call the function. Code files (libraries), either compiled or not, can be accumulated and included in a program. For example, many scientific applications, such as the Monte Carlo procedure used as the main example in this book, require calculation of the pairwise particle energy. Of course, the Lennard-Jones energy is only one pair energy example; many others exist. A library of pair energy functions written, debugged and validated once, would be a long-term time saver.

The primary caution here refers back to Table 1.1: "The Scientist Programmer tends to favor short development time over performance." To use libraries, function calling is necessary, and parameter passing (rather than the use of GLOBALS) is the norm. The point is that by purchasing or downloading the latest greatest function library, one will get a program up-and-running quickly, since you need only supply wrappers to the function calls. The downside is that such a program will not, no matter how optimized the library, execute as fast as possible.

Many programmers continue to debate the issue of superiority between Fortran and C/C++. In part, this is due to Fortran's amenability to the Scientist Style. Specifically, Fortran uses highly optimized libraries for common (and not-so-common) functions. Can a program that `CALLS` highly optimized library functions still be inefficient, even if the compiler option "In Line Functions" is used?

It may seem that since modern compilers can compile to "In Line Functions" that the programmer is getting the best of both worlds – modularity and fast execution. This compiler option allows the programmer to write source code in the Subroutine Style. At compile time, the compiler, *if it can,* generates explicit 'In Line' object code in place of the `CALL` statements. Most compilers will only substitute inline code if the function is small. Further, not all compilers offer an "inline functions" option, so that if the inline advantages are to be gained, the code must be hand written. The discussion in this chapter focuses on using inline code when the function is called many times, regardless of the size of the function. Using the compiler Inline option compared to code that does the `CALL`ing, will be much faster; however, there exists a potential limitation.

Often, libraries consist of functions that are written 'generically,' which may be considered an extension of the Scientist Style (recall from Chapter 1 that the Scientist Programmer tends to use complicated code to perform a simple task, often in the interest of shorter source code). Consider a routine in the library that manipulates a square (n x n) matrix. The library may contain a separate routine for a 2 x 2 matrix, a 3 x 3 matrix, a 4 x 4 matrix, etc. This could make the library quite large with repetitious code. On the other hand, the library function may be written with a variable matrix size, so the dimension is passed along with pointer(s) to the matrix data. The performance considerations of 'generic' vs. explicit functions of this type are considered Chapter 7.

Modularity is important in regard to updating and improving distributed applications. This is less true for 'one-off' applications running on a single machine in a research laboratory. However, even in this case, function calling to modular libraries may provide a valuable debugging or logic testing advantage. If an error is found in a function, it is far easier to make a change in one place (either subroutine or library) compared to hunting down the error in each place the function is used 'in-line.' Large programs are much more easily updated by modifying libraries, and many applications can benefit from simple updates if shared libraries (such as dlls) are used.

The benefits of using libraries and other modular source code are typically taken to outweigh the few percent of performance degradation they cause. This is not always the case. Any function calling, including calls to libraries, involves wasted CPU cycles – time the CPU is performing no useful work toward the actual calculation. This cost is large in iterative procedures as many modeling and simulation programs are. In a sense, one is trading CPU cycles in computation at run-time for 'human' cycles at design time. This may be exacerbated by the built-in generality of the library function. Demo *Function06* illustrates these points.

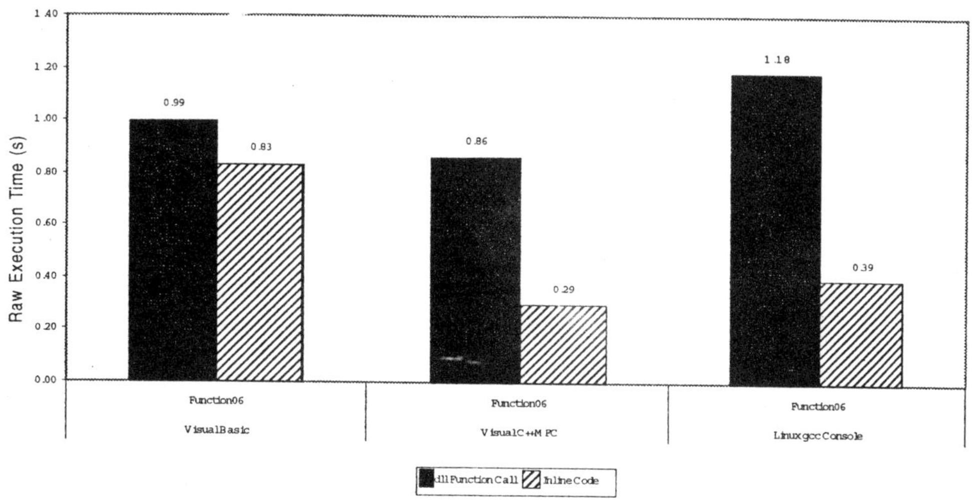

Figure 6.7: Raw execution times (Demo *Function06*) for 10,000,000 iterations for a library function call. A function call (passing two values) to an optimized dll (Windows) or so (Linux) routine is compared inline code.

To prepare the *Function06* demo, a Dynamic Linked Library (dll) was created (for the Linux demo, an SO, 'shared object' library was used). For this demonstration, the dll was written in VC++ using compiler optimizations (and is given as the *NumDemo* workspace in the "Chapter Demos/C/Windows/Function06" folder on the CD-Rom). It is therefore fair to say this simulates using an optimized library function. The dll contains only one exported function, a Lennard Jones Energy function. In the *Function06* code, a call is made to the LJ_Energy function in the dll. The pair energy is thus computed by passing the function the separation distance and the $Sigma^2$ value.

Execution data for using the dll function call and the inline code are shown in Figure 6.7. In this case, the calling program was also implemented in Visual Basic, a language often assumed to be too slow for real computational performance (the Visual Basic program is calling the *exact* same dll, written and optimized in C, as the C version). A comparison of BASIC to C in the demo is instructive in this case. It is important to emphasize that the *inline BASIC code is faster than calling the OPTIMIZED C library function, even from a C program.* This is due entirely to the cost of *calling* the function in the library. While the modularity and reuse arguments of library functions are clear counter-advantages for the inline technique, when performance is the primary goal, inline code is preferred. Often, the inline code should be hand written (or 'hand-rolled' as many programmers call it) for the highest performance.

6.5 HAND CODING VS. USING LANGUAGE FUNCTIONS – A PSEUDO RANDOM NUMBER GENERATOR

A similar argument can be made regarding the use of functions that many consider to be intrinsic to the programming language (though this may be due to library calls as discussed in Section 6.4). As an example, consider the `rand()` function, which of course must be called many times during a Monte Carlo simulation. This function returns an integer pseudo random number (between 0 and `RANDOM_MAX`) as computed using some generating function. Thus `RANDOM_MAX` defines the period of the generator (the number of pseudo random numbers generated before repeating the sequence).

One generating function for pseudo random numbers between 0 and 1 is

$$X_{k+1} = (a * X_k) \bmod m$$

$$X_k = X_{k+1}$$

where mod returns the remainder of $m/(a * X_k)$, a is an integer that defines the generator and the period of the function is given by m. The first value of X_k is the seed of the function. If the generator is used in a function that is used repeatedly, X_k should be declared as a global or *static* variable so the sequence is followed; otherwise, the same number will be generated. Using $a = 25$, $m = 2^{24}$ and a seed of 200, the demo *Function07* computes pseudo random numbers with this function. The performance data for *Function07* are shown in Figure 6.8. In this case, the hand coded

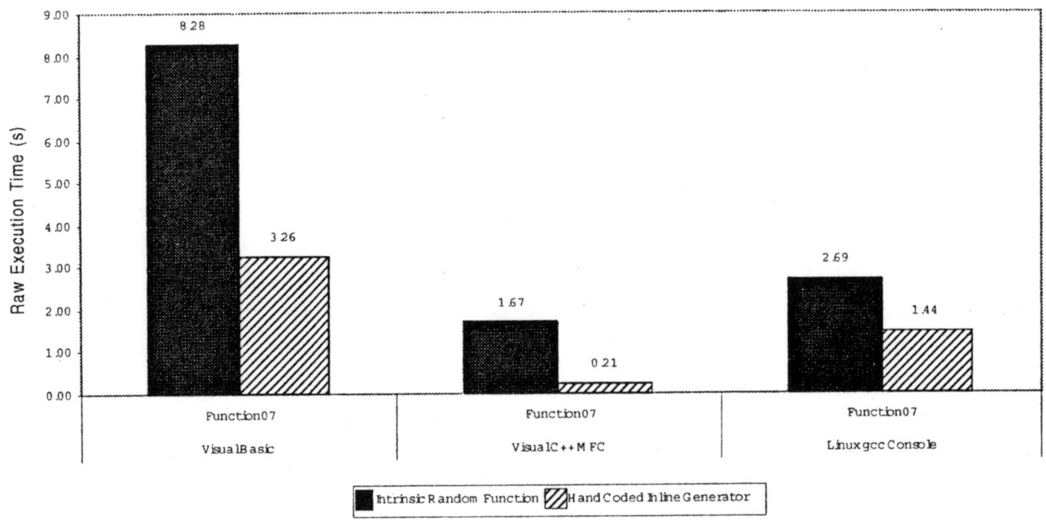

Figure 6.8: Raw execution times (Demo *Function07*) for 50,000,000 iterations a function call to a language function (actually a library function). The use of the RND() or RAND() functions are compared to an inline, hand coded pseudo random number generator.

('inline') pseudo random number generator is almost 8 times faster than the `rand()` standard library function. As such, a Monte Carlo simulation, or other computation requiring repetitive random number generation, will benefit from inline hand coding the pseudo random number function. A variation on this technique of eliminating language calls is explored in Chapter 8 where floating-point intensive intrinsic functions are eliminated.

6.6 RECURSIVE FUNCTION CALLING – THE FACTORIAL

If a programming language supports a function calling itself, the language is said to support 'recursive' functions. Many procedures are inherently recursive, and the programming of recursive procedures generates compact source code. However, recursive function calling generally involves considerable stack manipulation, and if the overall procedure is large enough, may even exceed the stack space of the program. The alternative to recursive function calling is using an iterative procedure.

Many recursive procedures may be difficult or inefficient to program iteratively, such as binary tree search algorithms. However, other procedures may benefit from iterative, rather than recursive execution. The benefits of iterative procedures are two-fold: they can be coded inline and they reduce (or eliminate) stack manipulations. An example of a common routine that can be coded either recursively or iteratively is the factorial function.

The factorial function is defined as n! = n * (n-1) * (n-2) * ... * 3 * 2 *1, and it is readily seen that this can be coded quite cleanly in a recursive function. Such a function is shown in **Listing 6.8**, and the iterative version is shown in **Listing 6.9**.

Listing 6.8 Recursive Factorial Function

```
double Recurs_Fact(int Arg1){

      double answer;

      if(Arg1==1) return(1);
      answer = Recurs_Fact(Arg1 -1) * Arg1;

      return(answer);

}
```

Listing 6.9 Iterative Factorial Function

```
double Iter_Fact(int factval){
```

```
        //integers
        int fact;        //factorial 'counter'

        //doubles
        double Result; //return value

        fact = 1;
         factorial = 1;

        //initialize Result
        Result = 1;

         //compute factorial iteratively
        for(fact = 1; fact<=factval; fact++){

             Result *= fact;

        }

        return(Result);
}
```

It is readily seen in **Listing 6.8** that as long as the parameter Arg1 > 1, the recursive function calls itself with Arg1-1. This pushes each value, and each return value, onto the stack. The iterative procedure, on the other hand, has no such stack manipulations. Empirical execution data for these codes, obtained from Demo *Function08*, are shown in Figure 6.9.

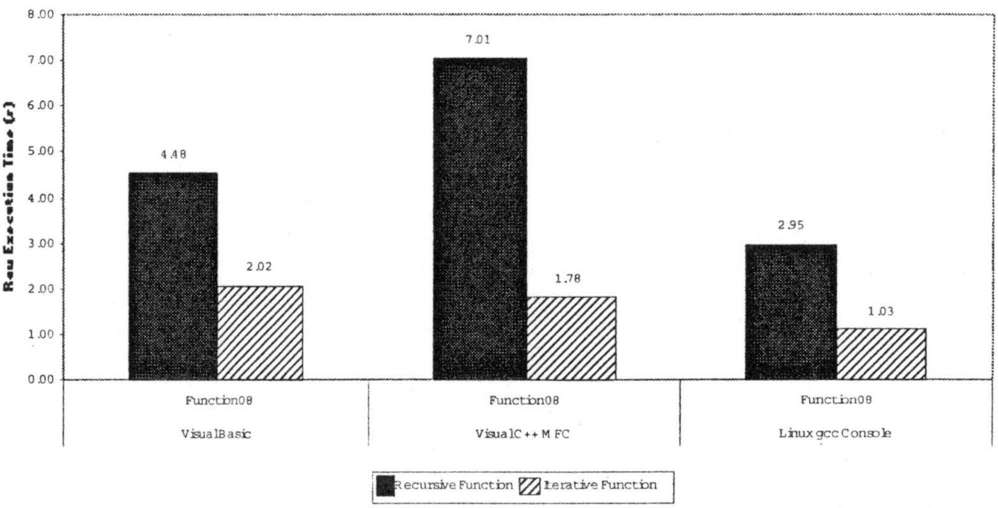

Figure 6.9: Raw execution times (Demo *Function08*) for 1,000,000 iterations for the calculation of 100!. A call to a recursive function is compared to the inline, iterative code.

6.7 FUNCTION CALLING CONVENTIONS

Another important consideration when programming function calls is the language specific calling convention. The calling convention refers to how data are placed on the stack, and which code module 'maintains' the stack. Two possibilities exist for each of these. The data can be placed 'left-to-right' or 'right-to-left' as they appear in the function call source code, and either the caller or callee can be responsible for stack clean-up (removal of local data) after the function call is complete. Other considerations in the calling convention are case specificity and how decorated function names are built.

C/C++ has two common calling conventions. The `_cdecl` calling convention has data placed right-to-left and stack clean-up done by the calling function. This is in direct contrast to the `_stdcall` (which is often called `WINAPI` since the windows.h header defines WINAPI as _stdcall; this convention used to be `_pascal` or -`_fortran`) convention in which parameters are passed right-to-left and stack clean-up is done by called function.

When writing a library, it is important to know the calling convention of the language that will be using the library. BASIC, Fortran and many other languages use the `_stdcall` convention, so the exported functions should be declared in this way. Functions used exclusively by C programs can utilize -`_cdecl`, which is the default C convention.

There is another convention used in Microsoft C++: `_fastcall`. This convention uses parameter passing in cpu registers, if possible; the stack is not used. While this often result in faster code, the compiler output is difficult to predict and which registers are used is not defined in the convention.

Chapter 7: Loops and Vectors

It is easy to recognize the importance of vector operations in scientific applications. For example, many applications have a spatial component. In addition, other operations can sometimes be written to take advantage of vector techniques. However, the implementation of vector operations in code may or may not be high performance. In this chapter, some simple programming techniques for working with vectors are explored. The techniques focus on loop structures in code; therefore, these techniques are adaptable to any code requiring loops: matrices, tensors and iterative routines such as linear system solvers.

7.1 GENERAL VECTOR CONCEPTS AND THE POTENTIAL FOR COMPLEX CODE STRUCTURE

In computations, vectors are generally resolved into components. For example, the 2-dimensional spatial vector **a** may be written

Eq. (7.1) $$\mathbf{a} = a_x\hat{\mathbf{x}} + a_y\hat{\mathbf{y}}$$

As an alternative, the row or column matrix notation may be used, such as

Eq. (7.2) $$\mathbf{a} = \begin{bmatrix} a_x \\ a_y \end{bmatrix} \quad \text{or} \quad \mathbf{a} = \begin{bmatrix} a_x & a_y \end{bmatrix}$$

The row or column vector notation is especially convenient when considering spatial transformations, as required in graphical applications.

A common operation involving vectors is the dot product, which results in a scalar. The dot product, c, of two vectors **a** and **b** is defined as

Eq. (7.3) $$\mathbf{c} = a_x b_x + a_y b_y$$

This is easily coded as shown in **Listing 7.1**

Listing 7.1 **Simple vector dot product code**

```
c = a_x * b_x + a_y * b_y;
```

However, few Scientist Programmers would define the data this way (as individual variables, a_x, a_y, b_x and b_y) or write the code in this way for a large application. Specifically, the code in **Listing 7.1** is not 'general' with respect to dimensionality; for example, it cannot be used, as written, for a three-dimensional vector dot product.

The dot product can be generalized to vectors of any dimension by the use of arrays, with the index 'pointing' to the component. For example, the vector **a** above can be written in code as $a(1) = a_x$ and $a(2) = a_y$. Using arrays, the generalized dot product is written in the Scientist Style as shown in **Listing 7.2**.

Listing 7.2 **Generalized 2-d vector dot product using arrays**

```
// N is dimensionality; 2 in this case
N = 2;

c=0;

for(Index=1; Index <= N; ++Index){

    c += a[Index] * b[Index];

}
```

Listing 7.2 is very general; it is also short and puts complicated logic 'to work' for the programmer. Note that in a sense this is an exception to the Scientist Style of writing code as the equations appear on paper. Nonetheless, it remains an example of that style with respect to short, easy-to-type code. This generalized approach to computing the dot product does have two key performance costs: the indices (which in the compiled ASSEMBLY are offsets to a pointer pointing at the first array element) must be computed, or otherwise updated, each time through the loop and compute cycles are expended 'maintaining' the loop itself.

Maintenance of the indices will require at minimum incrementing the index register being used. Algorithms more complicated than a simple dot product (such as a Gauss-Seidel system solver) require computation of the index at each step. In addition, the loop itself requires at least one increment (of the loop counter), one comparison (of the loop counter to the exit value) and one conditional jump per iteration. Such loop overhead in the case of small loops with simple code *inside* the loop may easily exceed the computational effort of the working code. For a single 2-d vector dot product, the low performance of **Listing 7.2** is not noticed. On the other hand, if **Listing 7.2** is called millions of times in an application, the wasted cycles 'add-up' and an obvious performance hit is observed.

An improvement on **Listing 7.2** returns the code to the style of **Listing 7.1**, except that an array is used rather than the unique simple variables shown in Equation (7.3). This is shown in **Listing 7.3**, which is an example of a general optimization called 'unrolling the loop.'

Listing 7.3 **2-d vector dot product with the loop unrolled**

```
c = a[1] * b[1] + a[2] * b[2];
```

The potential improvement of **Listing 7.3** over **Listing 7.1** is subtle; in other places in the application, accessing the vector components as array elements may be more efficient.

Rolled-up loops have been observed in real computational source code. For example, a popular quantum chemistry package contains many Cartesian coordinate manipulations in rolled loops running from 1 to 3. Such loops are excellent candidates for loop unrolling: the loop size is fixed (ie, loop parameters are immediate data rather than variable data) and the loops are quite small. Loop unrolling can result in performance gains in a variety of contexts. For example, the simple array assignment in **Listing 1.3** executes faster than the 'rolled-up' version in **Listing 1.2**. In addition to the specific examples given here, other routines can benefit, such as a fixed size Gauss-Seidel solver. Results for the demo programs with rolled and unrolled codes are shown in Figures 7.1 and 7.2. Generally, any time the code *inside* the loop is faster than the loop overhead and the unrolled code can fit in the CPU instruction cache, unrolling is advantageous. Empirical absolute latencies per iteration, corrected for CPU clock frequency, are shown in Table 7.1.

7.2 LOOP UNROLLING IN A TRADE-OFF WITH GENERALITY – MATRIX MULTIPLICATION

The matrix multiplication routine in Sample Program *Loop03* illustrates an important concept that blends the topics of Chapter 6 with loop unrolling. When library routines contain a function or procedure for vector dot product, matrix multiplication or other operations whose 'compute space' can be variable, they are generally coded as rolled loops with the dimension passed as a parameter. This hurts performance two ways. First, as shown in Chapter 5, having a variable loop size (requiring a memory access) can be slower than using an explicit loop size (allowing an immediate operand).

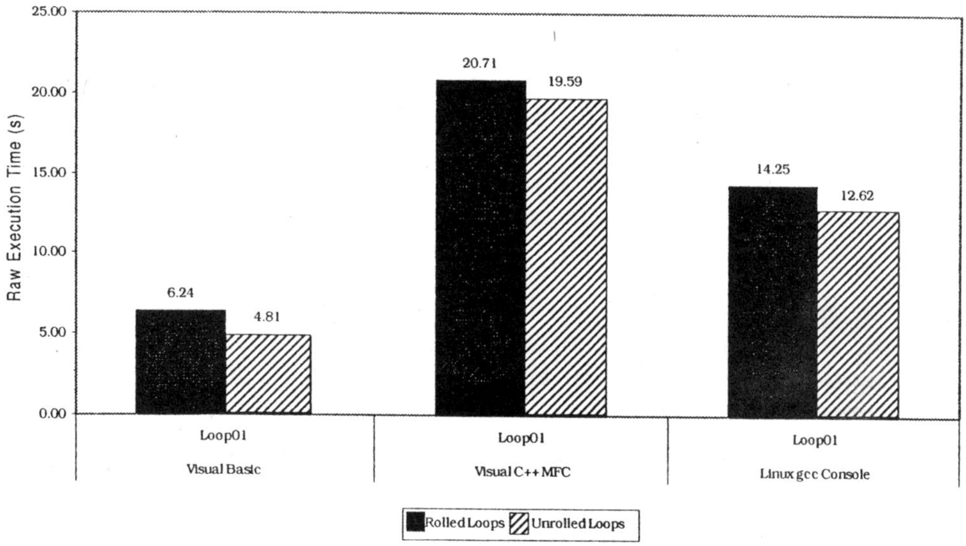

Figure 7.1: Demo *Loop01* raw execution times on the test computer for 10,000,000 iterations for a simple function assignment. Compared are the algorithms using rolled and unrolled loops.

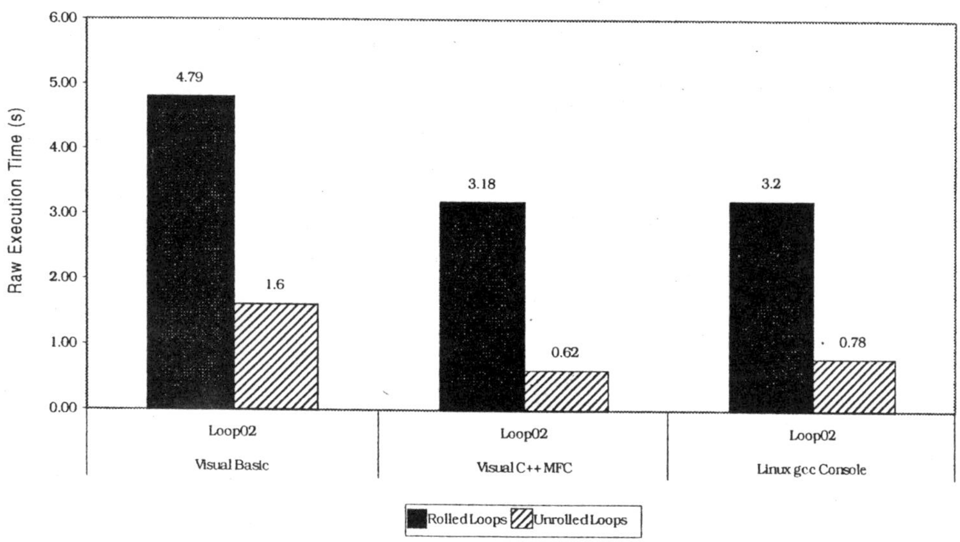

Figure 7.2: Demo *Loop02* raw execution times on the test computer for 100,000,000 iterations of a two dimensional vector dot product. Compared are the algorithms using rolled and unrolled loops.

Table 7.1: Reduced empirical clock cycles required per iteration for Rolled and Unrolled Loops.

	Rolled	**Unrolled**
Loop01	2485	2350
Loop02	38	7

Second, the loop is rolled-up, so loop overhead wastes cycles.

Unrolling the loops can only be done with explicit loop sizes. If the library is general, to be applied to many different computational spaces, the library must contain a separate procedure for each possible 'dimensionality.' In the broadest sense, this would be considered 'poor programming' since much code is somewhat redundant and code maintenance may be difficult. Further, if the calling application can use different dimensionality, the calling procedure must 'decide' which library procedure to call. For example, consider an application that performs n × n matrix multiplications, with n <= 8. The library must contain a procedure for 2 × 2, one for 3 × 3, etc, up to a procedure for 8 × 8. An even greater number of separate procedures are needed for non-square matrix multiplication, m × n. Here there are a large number of possibilities, and it must be remembered that matrix multiplication does not commute. (One approach to this 'assumes' that only column vectors of proper dimension are passed to a given routine, that is, rather than coding all possible $(m \times n)$ x $(n \times p)$ routines, the second matrix is composed of p column vectors and the routine is called repeatedly). For simplicity and illustration, only n × n matrix multiplication is considered here, and is shown in **Listing 7.4**

Listing 7.4 Generalized rolled-up n × n matrix multiplication

```
// N is dimensionality
for(I=1; I<=N; ++I){

   for(J=1; J<=N; ++J){

      c[I][J] = 0;

      for(k=1; k<=N; ++k){

         c[I][J] += a[I][k] * b[k][J]

      }

   }

}
```

This procedure works with matrices of 'any' size *N*. This code does not scale favorably since it contains three nested loops; **Listing 7.4** scales as N^3. For example, the code for 8 × 8 matrix multiplication will execute at least 8 times slower than the code for 4 × 4 matrix multiplication. The only way to improve this scalability is with parallel processing; this will be revisited in Ch. 12. In addition, the overhead for each successive loop multiplies the overhead, compounding the scaling problem. An explicit rolled-up version (shown for a 4 x 4 matrix) is shown in **Listing 7.5**.

Listing 7.5 **Explicit rolled-up 4x4 matrix multiplication**

```
for(I=1;I<=4; ++I){

   for(J=1;J<=4; ++J){

      C[I][J] = 0;

      for(k=1; k<=4; ++k){

         c[I][J] += a[I][k] * b[k][J]

      }

   }

}
```

In the absence of any caching or other advanced CPU functionality, **Listing 7.5** executes faster than **Listing 7.4** with *N*=4 since the loop parameters are immediate operands (rather than memory fetches of variable data). The unrolled version (for 4x4) is shown in **Listing 7.6**.

Listing 7.6 **Unrolled 4x4 matrix multiplication**

```
// I = 1
C[1][1] = A[1][1] * B[1][1] + A[1][2] *
          B[2][1] + A[1][3] * B[3][1] +
          A[1][4] * B[4][1];

C[1][2] = A[1][1] * B[1][2] + A[1][2] *
          B[2][2] + A[1][3] * B[3][2] +
          A[1][4] * B[4][2];

C[1][3] = A[1][1] * B[1][3] + A[1][2] *
```

```
              B[2][3]  +  A[1][3]  *  B[3][3]  +
              A[1][4]  *  B[4][3];

C[1][4]  =    A[1][1]  *  B[1][4]  +  A[1][2]  *
              B[2][4]  +  A[1][3]  *  B[3][4]  +
              A[1][4]  *  B[4][4];

// I = 2
C[2][1]  =    A[2][1]  *  B[1][1]  +  A[2][2]  *
              B[2][1]  +  A[2][3]  *  B[3][1]  +
              A[2][4]  *  B[4][1];

C[2][2]  =    A[2][1]  *  B[1][2]  +  A[2][2]  *
              B[2][2]  +  A[2][3]  *  B[3][2]  +
              A[2][4]  *  B[4][2];

C[2][3]  =    A[2][1]  *  B[1][3]  +  A[2][2]  *
              B[2][3]  +  A[2][3]  *  B[3][3]  +
              A[2][4]  *  B[4][3];

C[2][4]  =    A[2][1]  *  B[1][4]  +  A[2][2]  *
              B[2][4]  +  A[2][3]  *  B[3][4]  +
              A[2][4]  *  B[4][4];

// I=3
C[3][1]  =    A[3][1]  *  B[1][1]  +  A[3][2]  *
              B[2][1]  +  A[3][3]  *  B[3][1]  +
              A[3][4]  *  B[4][1];

C[3][2]  =    A[3][1]  *  B[1][2]  +  A[3][2]  *
              B[2][2]  +  A[3][3]  *  B[3][2]  +
              A[3][4]  *  B[4][2];

C[3][3]  =    A[3][1]  *  B[1][3]  +  A[3][2]  *
              B[2][3]  +  A[3][3]  *  B[3][3]  +
              A[3][4]  *  B[4][3];

C[3][4]  =    A[3][1]  *  B[1][4]  +  A[3][2]  *
              B[2][4]  +  A[3][3]  *  B[3][4]  +
              A[3][4]  *  B[4][4];

// I=4
C[4][1]  =    A[4][1]  *  B[1][1]  +  A[4][2]  *
              B[2][1]  +  A[4][3]  *  B[3][1]  +
              A[4][4]  *  B[4][1];
```

```
C[4][2]  =  A[4][1]  *  B[1][2]  +  A[4][2]  *
            B[2][2]  +  A[4][3]  *  B[3][2]  +
            A[4][4]  *  B[4][2];

C[4][3]  =  A[4][1]  *  B[1][3]  +  A[4][2]  *
            B[2][3]  +  A[4][3]  *  B[3][3]  +
            A[4][4]  *  B[4][3];

C[4][4]  =  A[4][1]  *  B[1][4]  +  A[4][2]  *
            B[2][4]  +  A[4][3]  *  B[3][4]  +
            A[4][4]  *  B[4][4];
```

Notice that **Listing 7.6** contains *no loops*, and therefore, no loop overhead. All CPU cycles work toward solving the problem. There is, however, a potential overall cost in this approach that must be examined before the style of **Listing 7.6** is chosen over **Listing 7.4**.

The generalized, rolled-up code in **Listing 7.4** has the advantage that it can be used in many applications requiring an *n* x *n* matrix multiplication. Indeed, only the dimensionality parameter need be passed to the procedure (assuming the arrays being used are globals; otherwise, pointers to the arrays would also have to be passed). The dimensionality can even be an immediate in the calling statement. More to the point, no decision logic is needed. On the other hand, for both **Listing 7.5** and **Listing 7.6**, a decision would need to be used in the calling application to make the proper call. This is emphasized since the decision logic overhead may outweigh the benefit of unrolling the loops; if so, it would be concluded that **Listing 7.4** is preferred.

In the test code Demo *Loop03*, this “stress” on the algorithm is simulated by randomly selecting either a 4 x 4 or 8 x 8 matrix multiplication (that is, about 50% each, but in random order). An array of random numbers is used so the same sequence of random numbers is used for each routine in the test. Further, the generalized rolled test is done with and without the decision structure.

The 4 x 4 unrolled code was typed in several minutes, which is not much more than either the rolled 4 x 4 code or the generalized version. However, the 8 x 8 unrolled code took longer, even using editing tools to copy and replace. Further, the 8 x 8 code would be very difficult to debug if only one array index is mistyped. Finally, the application with the 8 x 8 unrolled code took longer to compile, versus about one second for the code containing only rolled-up loops.

The execution times for each routine in Demo *Loop03* are shown in Figure 7.3 and empirical cycles shown in Table 7.2. This once again shows that run time performance is often gained as trade-off with development time. On the plus side, though unrolled code takes longer to type and compile, these operations need only be done ‘once’ per application. This may often be considered an acceptable compromise for code that executes approximately 10 times faster.

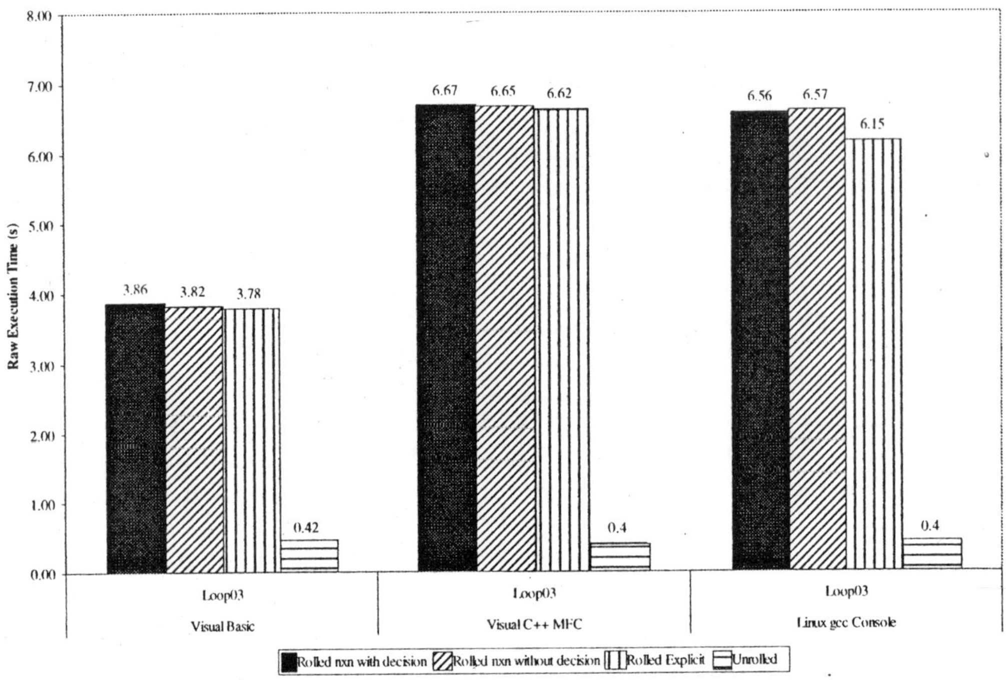

Figure 7.3: Demo *Loop03* raw execution times on the test computer for 500,000 iterations of nxn matrix multiplications. 4×4 and 8×8 matrices are multiplied in random order. Compared are algorithms using the generic nxn function with a decision structure in the calling routine, the generic nxn function with no decision structure in the calling routine, the explicit 4x4 and 8x8 functions with rolled-up loops and the explicit 4×4 and 8×8 functions with unrolled loops.

Table 7.2: Reduced empirical clock cycles required per iteration for several nxn matrix multiplication algorithms in Demo *Loop03*. This demo explores the overall issue of code reuse for a generic matrix routine versus the loss of such generality when using the explicit unrolled routines.

	Rolled	**Unrolled**
Generic with decision structure	15984	--
Generic without decision structure	15960	--
Explicit Loops	15888	960

7.3 PARTIAL LOOP UNROLLING – SIMPSON'S RULE INTEGRATION

Some mathematical procedures may not be readily amenable to completely unrolling the loops as shown in the previous examples. This may be due to the size of the loop being prohibitive to code explicitly. Further, even if the algorithm *can* easily be coded unrolled, the advantages of unrolling fade for loops larger than 10-20 (depending on the code inside the loop). The example chosen to illustrate that unrolling can be useful even in the case of a very large loop is Simpson's Rule numerical integration. Numerical integration is useful in a variety of modeling problems. For example, Simpson's Rule integration is one method used to evaluate the Debye Integral, which is of the form:

Eq. (7.4) $$D(x) = \frac{M}{x^M} \int_0^x \frac{x^M}{e^x - 1} dx$$

where M is an integer (1, 2 or 3). This integral is used for the statistical evaluation of several important thermodynamic properties of solids that are in turn important for the numerical modeling of a large variety of detonation phenomena. Numerical integration is also important in the statistical calculation of Second Virial Coefficients (used with the Virial Equation of State):

Eq. (7.5) $$B_2(T) = -2\pi \int_0^\infty [e^{-E(r)/(k_b T)} - 1] r^2 dr$$

In this equation, *E(r)* is a two-particle energy function, such as the Lennard Jones energy exemplified throughout this text. For simple energy functions (such as the hard sphere or square well functions), this integral can be evaluated analytically; however, inclusion of most realistic energy functions requires numerical integration.

Assuming the function f(x) can be integrated on the interval [a,b], the integral, I, can be calculated using Simpson's Rule as follows:

Eq. (7.6) $$I = \int_a^b f(x) dx$$

Eq. (7.7) $$I = \frac{h}{3}(s_0 + 2s_1 + 4s_2)$$

where

Eq. (7.8) $$h = \frac{b-a}{2n}$$

Eq. (7.9) $$s_0 = f(a) + f(b)$$

Eq. (7.10) $$s_1 = f(a+h) + f(a+3h) + \ldots + f(b-h)$$

Eq. (7.11) $$s_2 = f(a+2h) + f(a+4h) + \ldots + f(b-2h)$$

To use Simpson's Rule, the interval [a,b] is divided into an *even* number of 'subintervals,' with the length of each subinterval = h. To code Simpson's Rule, the index is 'slipped' so $f(a)$ is the first element of the function array in memory:

Eq. (7.12)
$$\begin{aligned} g_0 &= g(0) = f(a) \\ g_1 &= g(1) = f(a+h) \\ g_2 &= g(2) = f(a+2h) \\ &\vdots \\ g_{2n-2} &= g(2n-2) = f(b-2h) \\ g_{2n-1} &= g(2n-1) = f(b-h) \\ g_{2n} &= g(2n) = f(b) \end{aligned}$$

and by substitution,

Eq. (7.13)
$$\begin{aligned} s_0 &= g_0 + g_{2n} \\ s_1 &= g_1 + g_3 + g_5 + \ldots + g_{2n-1} \\ s_2 &= g_2 + g_4 + g_6 + \ldots + g_{2n-2} \end{aligned}$$

Listing 7.7 illustrates the rolled-up code, using N=2n.

Listing 7.7 **Simpson's Rule integration using rolled-up loops**

```
h_3    =  (b-a)/3N;
s0     =g(0)  +  g(N-1);
s1     =0;
s2     =0;

for(J=1;  J<=N-2;  ++J){
```

```
        s1  +=  g[J];
        s2  +=  g[J+1];



s1  *=  2;
s2  *=  4;

I  =  h_3  *  (s0  +  s1  +  s2);
```

The demo code *Loop04* uses Simpson's Rule to integrate the function x^5 on the interval [0, 0.9999] using 10,000 subintervals. The integral, I, is computed 50,000 times to simulate a 'larger' computational problem requiring iterative integration. For example, in dressed state modeling and other time dependent quantum mechanics calculations, the Schrodinger Equation is integrated numerically. Clearly, with 10,000 subintervals, complete unrolling of the loop is not feasible. Are we stuck with the performance of the rolled-up code in **Listing 7.7**?

A key optimization with a significant performance gain can be made to **Listing 7.7** by *partially* unrolling the loop. With partial unrolling, the loop step size is determined by the degree of unrolling the loop. One thing to consider is that the number of terms being summed may not be evenly divisible by the step size used. Therefore, with partial unrolling, the loop must be 'ended early' and some clean-up code used after the loop. For example, the variable *LoopBreak* can be defined as

Eq. (7.14) $$LoopBreak = \left[\operatorname{int}\left(\frac{N-2}{STEP} \right) * STEP \right] + 1$$

From this value, the ending loop value can be calculated as

Eq. (7.15) $$LoopEnd = LoopBreak - STEP$$

Listing 7.8 shows the first step in the process of improving the loop from **Listing 7.7**. In **Listing 7.8**, *LoopBreak=9997*, which gives 9993 for *LoopEnd*. This allows the maximum amount of the loop to be unrolled, but leaves two terms not in the sum. These are included in some 'clean-up' code after the partially unrolled loop.

Listing 7.8 **Simpson's Rule integration loop partially unrolled**

```
for(J=1; J<=9993; J+=4){

       s1  +=  g[J]  +  g[J+2];
       s2  +=  g[J+1]  +  g[J+3];

}

// clean up extra terms
s1 += g[9997];
s2 += g[9998];
```

Listing 7.8 eliminates ½ of the counter increment and comparison loop overhead. (notice the loop is stepped by 4 rather than 2 as used in **Listing 7.7**). This can easily be extended to **Listing 7.9**, which has the loop unrolled a little further.

Listing 7.9 **Simpson's Rule integration loop further unrolled**

```
for(J=1; J<=N-2; J+=6){

       s1  +=  g[J]  +  g[J+2]  +  g[J+4];
       s2  +=  g[J+1]  +  g[J+3]  g[J+5];

}

// clean up extra terms

s1 += g(9997);
s2 += g(9998);
```

This code has 1/3 of the counter increments and comparisons in loop overhead compared to the rolled-up version in **Listing 7.7**. Such partial unrolling can be extended further until the loop is optimized.

Figure 7.4 shows the execution time vs. 'degree of unrolling' for the Simpson's Rule problem serving as the present example and a sample absolute latency per iteration is shown in Table 7.3. Here, the 'degree of unrolling' is defined as the number of explicit terms for s1 or s2 inside the summation loop. Note that a diminishing return, and even a performance degradation, is reached with a certain degree of loop unrolling. This occurs for several reasons, notably the size of the CPU instruction cache. Once the sizes of the CPU caches are reached, performance degrades since the hardware advantages of caching prefetched and decoded instructions are lost. Since different CPU's have different cache characteristics, the performance plateau reached with loop

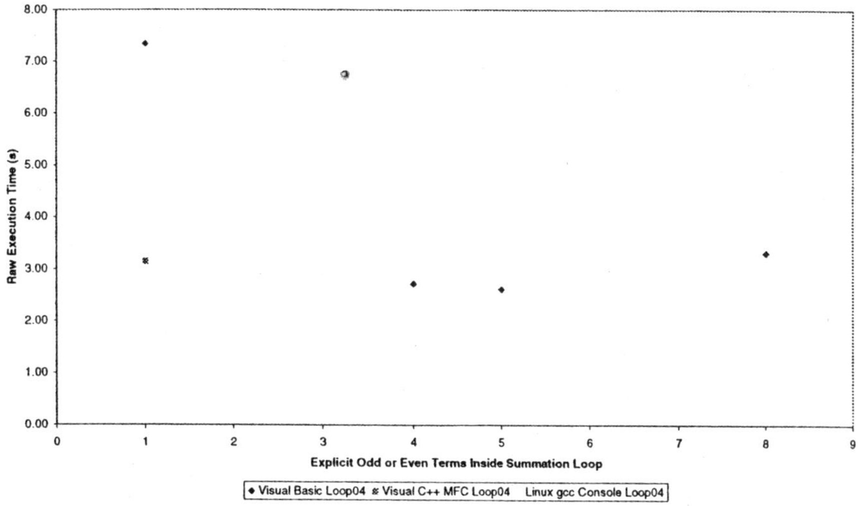

Figure 7.4: Raw execution time versus Number of Terms in Summation Loop for 50,000 iterations of a Simpson's Rule numerical integration. The number of terms is essentially the degree to which the loop is partially unrolled. These data were computed using Demo *Loop04*.

Table 7.3: Reduced empirical clock cycles required per iteration for rolled and several partially unrolled numerical integration algorithms in Sample program Loop04.

	Rolled	**Unrolled 4 Terms**	**Unrolled 5 Terms**	**Unrolled 8 Terms**
Visual C++	75,360	38,160	36,240	34,560

unrolling again emphasizes that broad portability and architecture-general code is not truly high performance.

The degree of unrolling can be computed and assigned dynamically for a general loop size (rather than the 'known' loop size of 10000 as given in the example) function. In the present example of Simpson's Rule integration, the *LoopBreak* and *LoopEnd* terms are listed above. Code for general Simpson's Rule, unrolled four terms in the summation loop, is shown in **Listing 7.10**. For general code, however, the clean-up terms cannot be stated explicitly (unrolled); notice that this introduces a performance degradation in the form of a very small rolled-up loop. This further emphasizes the notions mentioned in several earlier chapters that code generality limits performance.

Listing 7.10 General Partially Unrolled Simpson's Rule Loop

```
LoopBreak = (int(N-2)/8) * 8) + 1;

LoopEnd = LoopBreak - STEP;
'main summation loop
for(J=1; J<=LoopEnd; J+=8){

      s1 += g[J] + g[J+2] + g[J+4] + g[J+6];
      s2 += g[J+1] + g[J+3] g[J+5] + g[j+7];

}

// clean-up extra terms
for(J=LoopBreak; J<=N-2; J=J+2){

      s1 += g(J);
      s2 += g(J+1);

}

// continue with algorithm
```

Numerous procedures are amenable to the technique of partial unrolling. Like general loop unrolling, performance gains are observed when operations inside the loop are relatively simple. Further improvements to the mathematical procedures discussed in this chapter can be made. Specifically, many vector operations are suitable candidates for parallel operations. These will be discussed further in Chapter 12.

7.4 A PRACTICAL EXAMPLE: TIME DEPENDENT HEAT FLOW

Many of the examples in this book are artificial demonstrations; the algorithm of interest is wrapped with an iteration loop to simulate a larger computational problem. To illustrate loop unrolling in a real, though small, example, consider the one – dimensional time dependent heat flow of a thin plate as modeled using the Finite Volume Method. The problem is adapted from Chapter 8 of *Computational Fluid Dynamics: The Finite Volume Method.*

Figure 7.5 shows the basic problem and the uniform mesh used in this calculation. The 2.0 cm plate is divided into five mesh points, each 0.004 m apart. The plate is initially ($t < 0$ s) at uniform temperature 200 °C. At $t = 0$, the right side of the plate is dropped to 0 °C (the other end is insulated, so heat flow there is zero). The problem

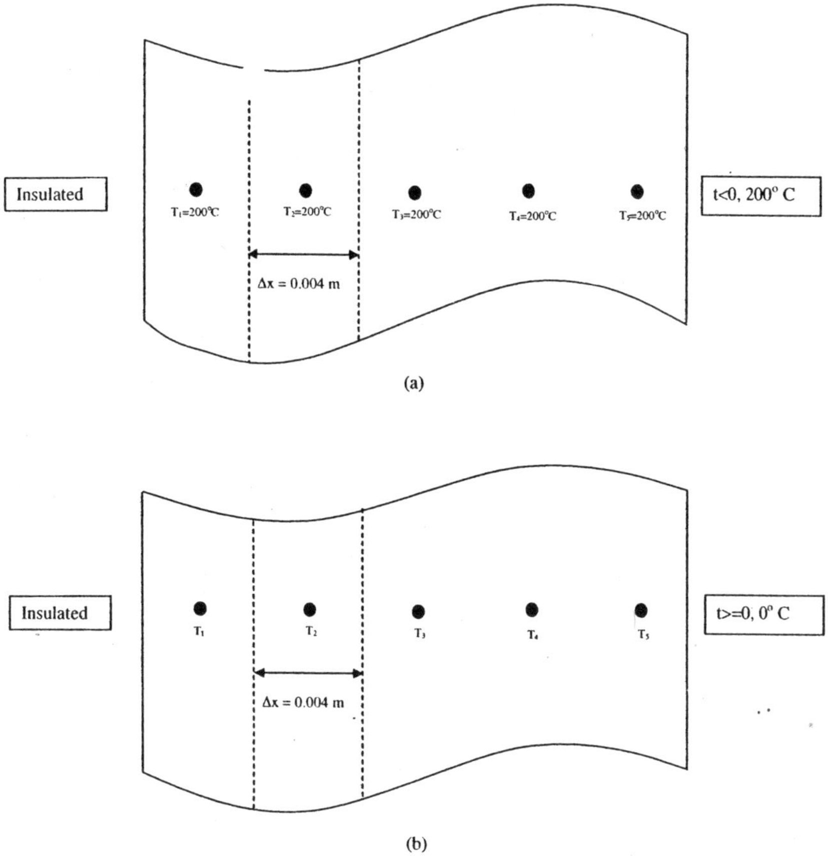

Figure 7.5: Schematic representation of the thin plate dynamic heat flow example modeled using the Finite Volume Method. Figure 7.5(a) shows the initial condition of the plate, at thermal equilibrium, before "t_0." At t_0, the temperature of the right side boundary is dropped to 0 °C. The plate is a sheet 2 cm along x and 'infinite' in the y direction.

is to compute the temperatures at each mesh point as a function of time, from 0 to 1200 seconds in 0.001 second intervals.

The transient heat conduction equation in one dimension is

Eq. (7.16) $$\rho c \frac{\partial T}{\partial t} = \frac{\partial}{\partial x}\left(k \frac{\partial T}{\partial x}\right)$$

for density ρ, specific heat c, temperature T, time t, position x and thermal conductivity k. For this sample problem, $\rho c = 1\times10^7$ J/(m^3 K) and $k = 10$ W/(m K). The Finite Volume Method uses the integrated form of the governing equation, which is then discretized onto the finite mesh. Such treatment yields the following system of five equations for the five unknown node Temperatures for a given time interval:

For Node 1: $400025\ T_1 = 25\ T_2 + 400000\ T_1^0$
For Node 2: $400050\ T_2 = 25\ T_1 + 25\ T_3 + 400000\ T_2^0$
For Node 3: $400050\ T_3 = 25\ T_2 + 25\ T_4 + 400000\ T_3^0$
For Node 4: $400050\ T_4 = 25\ T_3 + 25\ T_5 + 400000\ T_4^0$
For Node 5: $400075\ T_5 = 25\ T_4 + 400000\ T_5^0 + 50\ T_B$

Here a superscript ^{0} denotes the 'known' temperature at the beginning of the time step, and T_B is the right hand boundary Temperature. T_B for this demo is zero. T_x^0 represents the temperatures at the beginning of *each* time step, not specifically at time = 0; as the time is advanced, the T_x^0 are updated. For each time, the system must be solved for T_x. A generic procedure for solving the time dependent problem at sequential time steps is shown in Figure 7.6.

The actual system of equations can be written in matrix form,

Eq. (7.17)
$$\begin{bmatrix} 400025 & -25 & 0 & 0 & 0 \\ -25 & 400050 & -25 & 0 & 0 \\ 0 & -25 & 400050 & -25 & 0 \\ 0 & 0 & -25 & 400050 & -25 \\ 0 & 0 & 0 & -25 & 400075 \end{bmatrix} \begin{bmatrix} T_1 \\ T_2 \\ T_3 \\ T_4 \\ T_5 \end{bmatrix} = \begin{bmatrix} 400000T_1^0 \\ 400000T_2^0 \\ 400000T_3^0 \\ 400000T_4^0 \\ 400000T_5^0 \end{bmatrix}$$

or, in matrix-vector notation,

Eq. (7.18)
$$A\hat{T} = b$$

There are numerous algorithms for solving linear systems like the one in Eq. (7.17). Of particular note are ones that take advantage of the fact that the coefficient matrix, **A**, is tridiagonal. However, in many flow modeling calculations, particularly ones that give reasonable accuracy when modeling both convection and diffusion, the resulting matrix is *not* tridiagonal. A general method is Gauss-Seidel iteration. Gauss-Seidel iteration is used to solve the system of equations in this example, and the procedure as applied to Eq (7.17) is shown in Figure 7.7.

Demo *Loop05* executes the procedure summarized in Figure 7.6 using two versions: one with the loops rolled-up, and one with the loops unrolled. Note that several steps in the procedure summarized in Figure 7.6 can be unrolled: computation of the **b**,

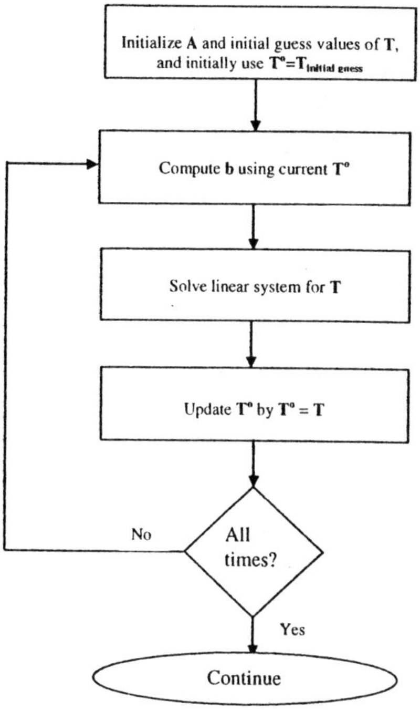

Figure 7.6: Outline of the numerical time dependent heat conduction solution. ***T*** is the node Temperature vector, ***T°*** is the vector for the node Temperatures at the start of each time step, ***A*** is the coefficient matrix and ***b*** is the right side (constant term) vector. Note that ***b*** and ***A*** are in part determined by size of the time step.

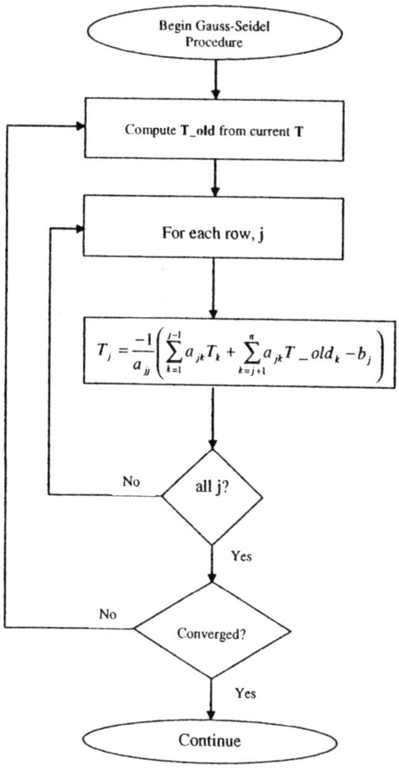

Figure 7.7: Summary of the Gauss-Seidel procedure used to solve a system of n linear equations, given as used for the one dimensional time dependent heat conduction problem. ***T*** is the node Temperature vector (with elements T_x), ***T_old*** is the node Temperature vector from the *previous* iteration, *j* is the row counter and *k* is the column counter. a_{nm} are the elements of the coefficient matrix ***A***. Not shown explicitly in the Figure are the updating of the iteration counter used to prevent an infinite loop in the case of non-convergence and calculations to test for convergence.

T_old and **T°** vectors, as well as the iteration loops of rows and columns themselves in the Gauss-Seidel part of the procedure. Raw execution data for *Loop05* are shown in Figure 7.8. Clearly, in this practical example, the unrolled coding of the algorithm executed 2 times faster for the C++ code. In other words, by taking the time to unroll the loops, problems two times larger (or twice as detailed) can be solved with the same computational effort.

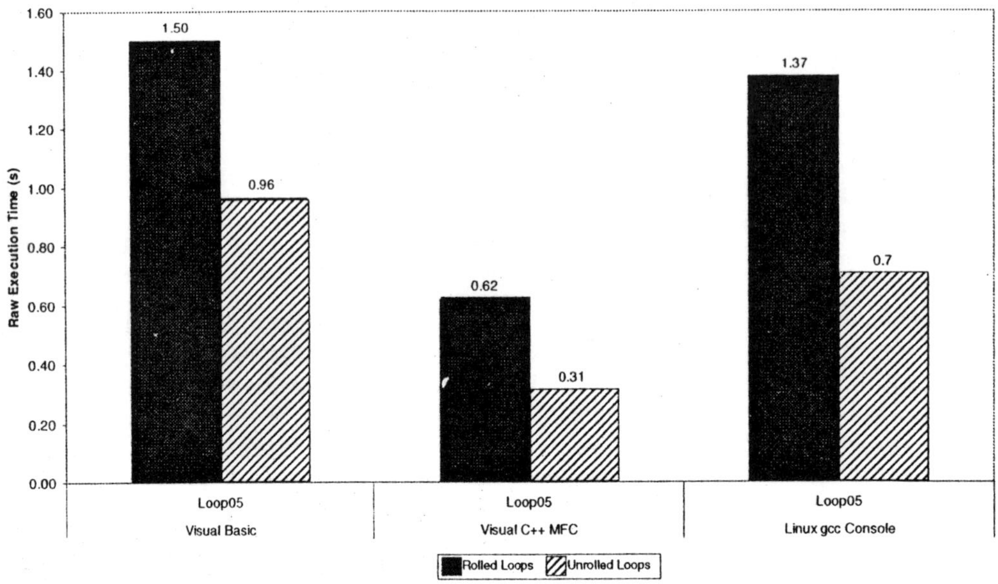

Figure 7.8: Demo *Loop05* raw execution times for the rolled and unrolled Gauss-Seidel solutions of the linear system that models the dynamic heat flow problem depicted in Figure 7.5. The system is solved for times 0.001 –1200 seconds at 0.001 second time steps, resulting in 1,200,000 solutions for each of the five node points.

Chapter 8: Programming in the RISC Style

In this chapter, the fundamental essence of the Scientist Programming style is examined. In Chapter 1, the position was presented that Scientist Programmers tend to write code that appears similar to equations on paper; the code is more readable by humans than computers. The flaws in this approach are presented in Chapter 8.

Chapters 8, 9 and 10 constitute the 'big leaps' in performance that can be gained with a single node, serial processing CPU. Algorithms are shown that execute ten or more times faster than the corresponding Scientist Programmer versions. Combined with the techniques of Chapters 5-7, the noticeable speed gains that are needed for robust computational modeling take shape.

8.1 THE KISS PRINCIPLE

Practicing scientists have generally all encountered the "Keep It Simple, Stupid," or KISS Principle. Experimental designs, theoretical explanations and modeling equations all benefit from adherence to this guideline. At worse, complex units are constructed from simple logical units.

The idea behind KISS is not that scientists are stupid and cannot function with complex ideas. Quite the contrary is true, but one might say that to eat a large plate of food, small bites are necessary. The same statements are true with execution of computer programs. The compilers *can* generate proper code from complex source structures, and of course, the CPU *can* execute the resulting object code. However, the issue of efficiency again arises.

Computers are fundamentally simple machines. After all, they 'understand' a language built from only two characters: 1 and 0. The high level languages used to write useful programs arise only from putting many simple pieces together. As an example, consider that in BASIC, C/C++ or virtually any modern language, scientists freely write exponentiation code, y^x (which is pow(y,x) when using the C math library). However, there is no base CPU instruction, or opcode, to perform this function in the general number base. The compiler must generate code using the $\log_2$ or 2^x capabilities of the floating-point processor. Without thought to 'perhaps there is a better way,' the Scientist Programmer writes the code in any convenient fashion the compiler will accept. Further, examination of many scientist source codes reveals that some programmers seem to take pride in how many floating-point operations they can 'cram' into a single line of source code. Recall, again, the two defining features of the Scientist Programmer: favors *short* source code, and favors code that looks like the math on paper.

There is a serious consequence to this style of programming. Modern processors are fundamentally RISC machines – Reduced Instruction Set Computers. A true RISC chip *requires* simple, 'bite-size' instructions and has no internal provision for inherent complexity in the instruction set, at least in the core execution units. An example of a true RISC chip might be one that has no integer multiply instruction, so that multiplies must be done by looped adding.

Of course, two selling points of the modern PC processors are that they are CISC (Complex Instruction Set Computers) and that they contain powerful floating-point processors. Recall from Chapter 2 that many modern processors are built around RISC cores. These chips have CISC 'preprocessor' circuitry that translate the CISC op-codes into RISC micro-operations that the ALU or FPU execution units can actually process. Even with these preprocessors, an important General Statement can be made, and will be developed throughout this chapter:

> *A CPU, even one with CISC capability, will execute RISC style algorithms faster than CISC ones.*

This leads to a key optimization: *Program in a RISC style.* If the programmer supplies "RISC style" high-level code, CPU effort will not be 'wasted' executing inefficient opcodes generated by the compiler. Such inefficiency may result from unnecessary library calls (Chapter 6), by doing certain simple calculations 'the hard way,' or by performing unnecessary operations.

8.2 PRACTICAL EXAMPLE – THE LENNARD-JONES ENERGY

The pair energy calculation that has been the focus of the optimizations in early chapters will again serve as the RISC style example. Recall the basic algorithm as shown in **Listing 8.1**.

Listing 8.1 Basic Scientist Style LJ Pair Energy

```
// Depth is well depth
// SIGMA is related to well position

// Depth, Num_Particles and SIGMA are defined

LJ_Energy = 0;
```

```
for(I=1; I<=Num_Particles-1; ++I){

   for(J=I+1; J<=Num_Particles; ++J){

      r = sqrt(pow((x[I]-x[J],2) + pow((y[I]- y[J]),2));

      LJ_Energy += pow(SIGMA/r),12) -
            pow((SIGMA/r),6);

   }

}

LJ_Energy *= C;
```

The two lines inside the innermost loop are very complicated. In fact, it has often been seen that these two lines are even combined into one line of code. Putting the *r* calculation in the actual energy calculation would be particularly bad because that would require *r* to be computed twice for each particle pair. Even as written, the compiler has to work hard at parsing this line correctly. Due to generalized parsing rules, it is reasonable to assume that a poor, inefficient op-code sequence results from the source code shown in Listing 8.1.

In general, the simpler an equation can be made, the faster the object code created by the compiler. There is a point of diminishing returns, however, when data transfers become excessive. Each algorithm, indeed each *calculation*, will have an optimum point in the trade-off between fast execution of op-code elements and data transfers of the cached data. Further, the optimal scheduling of the micro-operations by the CPU will be very context dependent. These statements emphasize the point that often *each calculation* must be optimized and generalized libraries, though optimized for the general context, are not always the fastest code for a particular application.

The pair energy calculation, removed from its context (that is, no sigma or depth parameters), serves as an example of the RISC style. In the test code, *r* was fixed, so that only the different compilations of actual LJ_Energy computation are compared. The test code is shown in **Listing 8.2**.

Listing 8.2 **Scientist CISC style coding of simplified LJ Energy**

```
// just a value to use
r = 1.634;

// Scientist Style enters
// equation as written on paper
Energy = 1/pow(r,12) - 1/pow(r,6);
```

Listing 8.2 is clearly Scientist Style source code. The energy calculation 'looks like' the reduced Lennard-Jones equation on paper. This source does not 'ease' the data to the compiler and the compiler cannot typically generate RISC style (that is, relatively easily decoded and scheduled) object code. This code ignores the fact that

Eq. (8.1) $$r^{12} = (r^6)^2$$

so that in essence r^6 is computed twice.

Several variations, such as **Listing 8.3** and **8.4**, were tried in an attempt to optimize the RISC style coding of the algorithm in **Listing 8.2**.

Listing 8.3 **a RISC style version of Listing 8.2**

```
// just a value
r = 1.634;


'r6 = r⁶
r6 = r * r * r* r * r * r;

'r12 = (r⁶)² = r¹²
r12 = r6 * r6;

'Energy = 1/r¹² - 1/r⁶
Energy = 1/r12 - 1/r6;
```

Listing 8.4 **a RISC style variant of Listing 8.3**

```
' just a value to use
r = 1.634;


'r3 = r³
r3 = r * r * r;

'r6 = (r³)² = r⁶
r6 = r3 * r3;

'r12 = (r⁶)² = r¹²
r12 = r6 * r6;

'Energy = 1/¹² - 1/r⁶
Energy = 1/r12 - 1/r6;
```

Other algorithms were tested, but **Listing 8.3** was found empirically to execute the fastest. To understand the results and write other algorithms faster, the data transfers, floating point operations and integer operations should be counted. Though a detailed view of the compiled ASSEMBLY Language is considered below, an estimation of the CPU instructions is adequate to optimize the code.

Many C compilers can be used to illustrate the details of how the CISC and RISC style codes are different at the basic instruction level. By setting the compiler to output a listing file (with .asm extension for MS Visual C++ 6.0 Pro; for gcc, the –S command line switch is used), the compiled ASSEMBLY instructions for each line of C source can be examined. Such listings for C equivalents to **Listing 8.2** and **Listing 8.3** are shown in **Listing 8.5** and **Listing 8.6**, respectively.

Listing 8.5 MS VC++ ASSEMBLY Language output for C code similar to Listing 8.2

```
Energy = 1/pow(r,12) - 1/pow(r,6);

      push 1076363264        ; 40280000H
      push 0
      mov  eax, DWORD PTR _r$[ebp+4]
      push eax
      mov  ecx, DWORD PTR _r$[ebp]
      push ecx
      call _pow
      add  esp, 16                ; 00000010H
      fdivr QWORD PTR __real@3ff0000000000000
      fstp QWORD PTR -324+[ebp]
      push 1075314688        ; 40180000H
      push 0
      mov  edx, DWORD PTR _r$[ebp+4]
      push edx
      mov  eax, DWORD PTR _r$[ebp]
      push eax
      call _pow
      add  esp, 16                ; 00000010H
      fdivr QWORD PTR __real@3ff0000000000000
      fsubr QWORD PTR -324+[ebp]
      fstp QWORD PTR _Energy$[ebp]
```

An examination of this assembled single line of source code shows that for EACH iteration, the CPU is performing (ignoring the loop overhead):

4	PUSH register	for a total of	4	Pentium cycles
4	PUSH immediate	for a total of	4	Pentium cycles
2	CALL far (external)	for a total of	8	Pentium cycles

2	FDIVR 64 bit mem	for a total of	78	Pentium cycles
2	FSTP 64 bit mem	for a total of	2	Pentium cycles
1	FSUBR 64 bit mem	for a total of	3	Pentium cycles
2	ADD register, imm	for a total of	2	Pentium cycles
4	MOV reg, 32 bit mem	for a total of	4	Pentium cycles

or 105 cycles total without even considering the cycles used in the actual _pow function. To contrast, consider **Listing 8.6**, the corresponding ASSEMBLY Language output for **Listing 8.3**.

Listing 8.6 MS VC++ ASSEMBLY Language output for C code similar to Listing 8.3

```
r6 = r * r * r * r * r * r;

        fld   QWORD PTR  _r$[ebp]
        fmul  QWORD PTR  _r$[ebp]
        fmul  QWORD PTR  _r$[ebp]
        fmul  QWORD PTR  _r$[ebp]
        fmul  QWORD PTR  _r$[ebp]
        fmul  QWORD PTR  _r$[ebp]
        fstp  QWORD PTR  _r6$[ebp]

r12 = r6 * r6;

        fld   QWORD PTR  _r6$[ebp]
        fmul  QWORD PTR  _r6$[ebp]
        fstp  QWORD PTR  _r12$[ebp]

Energy = 1/r12 - 1/r6;

        fld   QWORD PTR  __real@3ff0000000000000
        fdiv  QWORD PTR  _r12$[ebp]
        fld   QWORD PTR  __real@3ff0000000000000
        fdiv  QWORD PTR  _r6$[ebp]
        fsubp ST(1), ST(0)
        fstp  QWORD PTR  _Energy$[ebp]
```

which performs

4	FLD 64 bit mem	for a total of	4	Pentium cycles
6	FMUL 64 bit mem	for a total of	18	Pentium cycles

2	FDIV 64 bit mem	for a total of	78	Pentium cycles
3	FSTP 64 bit mem	for a total of	3	Pentium cycles
1	FSUBP reg, reg	for a total of	3	Pentium cycles

for a total of 106 CPU cycles. Note, however, there are *no hidden cycles* in this computation as for the CALLs to the _pow library function in **Listing 8.5**. Empirical results for 10,000,000 iterations of the simplified LJ Energy (Demo *RISC01*) are shown in Figure 8.1; absolute latencies per iteration are shown in Table 8.1. Notice that at the ASSEMBLY language level, the code can be optimized further by eliminating the FSTP _r6 followed by a FLD _r6 (which is storing and re-loading the same value in two back-to-back instructions). ASSEMBLY Language optimization is formally beyond the scope of this book, but the interested reader may wish to consider code scheduling (pairing) and register use rather than using so many 64 bit memory transfers (even though the L1 cache can be used).

Also, notice how costly the floating-point division is. This is an important point in optimizing iterative procedures that is often overlooked by the Scientist Programmer. *Division* by the CPU should be avoided; instead, when possible, *the programmer* should compute the reciprocal and use a multiply in code. For example, notice in each of the C++ demo's on the CD-Rom that multiplication by 0.001 is used to convert milliseconds to seconds, rather than division by 1000. At least, when possible, divisions should be performed *once* outside loops. However, this option is not possible

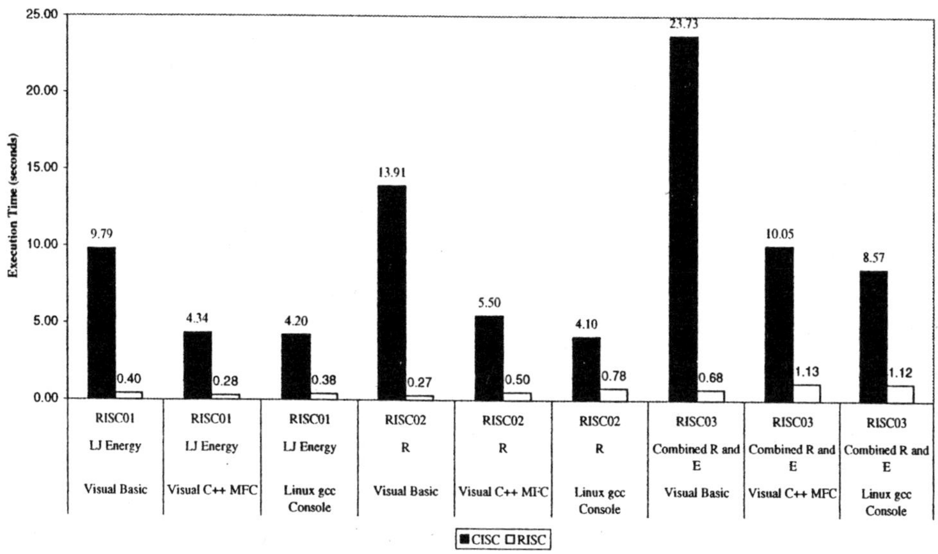

Figure 8.1: Demo *RISC01*, *RISC02* and *RISC03* raw execution times on the test computer for 10,000,000 iterations of the reduced Lennard-Jones energy pair energy calculation. Compared are the computations using a CISC style algorithm and using a RISC style algorithm. Shown are results for just the energy calculation, just the distance calculation and the combined distance and energy calculation.

Table 8.1: Reduced empirical clock cycles required per iteration for the CISC and RISC

	CISC	RISC
RISC01	516	29
RISC02	612	55
RISC03	1201	131
RISC04	238	275

in the Lennard Jones pair energy calculation; the computation forces the use of the costly floating-point division. Removing this CPU intensive division is addressed by the technique discussed in Chapter 9.

One may be wondering at this point that if the only real difference between the CISC and RISC styles in this demo is the function call (to `_pow`) itself, why is the performance difference so much greater than the function calling demo's in Chapter 6? In addition to the overhead of calling the function, one must consider *what the called function is doing*. In this case, the `pow(double base, double exponent)` is a *general* function, and can take *any* double precision floating point value as either base or exponent. Clearly, computing a number raised to a fractional exponent (which is possible for the general double precision value) is more costly than when using only integer exponents, as for the case of the Lennard Jones energy. In the case of small, integer exponents, the `pow()` function, which allows for any exponent, is simply 'overkill.' The result is wasted CPU effort. Even if the function contained 'simple' integer exponent code, such as that used for the RISC demo, there would be at least decision structure overhead and probably rolled loops as well. The use of generic, non-application specific library functions is therefore a Scientist Style approach designed specifically for rapid development at the expense of run-time performance.

Unfortunately, there is no magic reduction formula for all compilers. This also illustrates a disadvantage in using library packages: if the library functions are RISC optimized at all, it is likely the optimization is for a particular application, platform, compiler or hardware. The bottom line is that for high performance code, it must be decided where the line between development time and execution time is drawn. Only after rather detailed profiling of the code can the inefficiency of **Listing 8.2** be eliminated.

A similar optimization approach can be taken for the computation of the pair distance, *r*, in **Listing 8.1**. **Listing 8.7** and **Listing 8.8** show the Scientist Programmer and optimized styles, respectively.

Listing 8.7 **Scientist CISC style for computation of 2-d particle distance**

```
// some made-up values
// for the two coordinates
double x1=1.234;
```

```
double x2=2.675;
double y1=0.045;
double y2=2.073;

// 10,000,000 computations
// common in Monte Carlo procedures
for(counter=1; counter<=10000000; ++counter){

      // Scientist Style: equation as written
      r = sqrt(pow(x1 - x2,2) + pow(y1 - y2,2));

}
```

Listing 8.8 **RISC Style for computation of 2-d particle distance**

```
// some made-up values
// for the two coordinates
double x1=1.234;
double x2=2.675;
double y1=0.045;
double y2=2.073;

double TempX;
double TempY;

for(counter=1; counter<=10000000; ++counter){

            TempX = x1 - x2;
            TempX *= TempX;
            TempY = y1 - y2;
            TempY *= TempY;
            TempY += TempX;

            r = sqrt(TempY);

}
```

Execution times for **Listing 8.7** and **Listing 8.8** (Demo *RISC02*) are summarized in Figure 8.1 and Table 8.1. In addition, execution times for the combined *r* and LJ Energy calculation are also summarized in Figure 8.1 (Demo *RISC03*, which compares innermost code of **Listing 8.1** and a combination of **Listing 8.3** and **Listing 8.8**). Clearly, the RISC style constitutes a significant optimization of the Lennard-Jones pair energy computation; the RISC style in Demo *RISC03* executes nearly 10 times faster than the Scientist Style.

8.3 PRACTICAL EXAMPLE – THE BOLTZMANN FACTOR

The previous examples in this chapter dealt primarily with the floating-point computation of pair energies. The other main floating-point calculation in the Monte Carlo procedure is the Boltzmann weighting function. By pre-computing the argument to the `exp()` function, RISC style optimization can be applied to this calculation as well. However, since the x87 instruction set contains instructions related to 2^x (which can be used to calculate *any* constant^x), the RISC style causes unnecessary data moves. In other words, one might say that the exponential function is hardware optimized and needs no further software tricks to execute faster. The actual results for `exp()` CISC versus RISC, obtained from Demo *RISC04*, are shown in Figure 8.2 and Table 8.2.

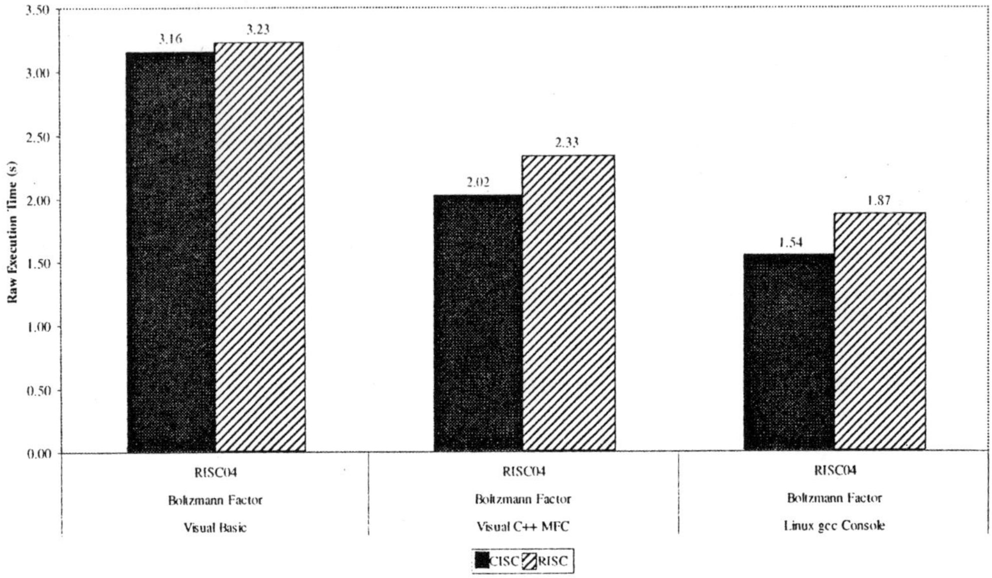

Figure 8.2: Demo *RISC04* raw execution times on the test computer for 10,000,000 iterations of the Boltzmann factor calculation. Compared are the computations using a CISC style algorithm and using a RISC style algorithm.

8.4 OTHER MATHEMATICAL TRICKS

Listing 8.1 was mentioned to have a poor structure not only in terms of being a CISC algorithm, but also because it contained a duplicate floating point computation. One optimization that results from careful profiling is to remove unneeded operations from the algorithms. For example, if an equation can be transformed to a larger "space," some floating-point math may be eliminated.

For example, suppose in the calculation all the particles within a certain distance of a reference point need be counted (such as the for the Probability Distribution Function calculation in the Monte Carlo procedure, but the reference point may be a reactive interface or any general point of interest in a given model). As an illustration, suppose the particles within 20 angstroms of the reference point are to be counted, and the Scientist Programmer code in **Listing 8.9** is the starting point (shown including the optimized distance calculation from **Listing 8.8**).

Listing 8.9 **a Scientist Style Radial Distribution Function calculation**

```
Within_Distance = 0;
for(Particle_Counter=1;
Particle_Counter<=Number_of_Particles;
     ++Particle_Counter){


     // compute distance between current particle
     // and reference particle

     tempX = X_reference - X(Particle_Counter);
     tempX *= tempX;

     tempY = Y_reference - Y(Particle_Counter);
     tempY *= tempY;

     tempY += tempY;

     Distance = sqrt(tempY);

     // count particle if within interesting
     // distance, which is 20 for this
     // example
     if (Distance <= 20) {
          Within_Distance += 1;
     }

}
```

Note that the distance is precomputed before the `if()` statement, so at least a CISC style calculation is not done in the `if()` line. The optimized distance calculation from **Listing 8.8** is employed in **Listing 8.9**; it is important to emphasize that high performance code results when the optimizations build upon each other.

The optimization point to make from **Listing 8.9** is in asking "why is the square root even calculated?" The `sqrt` function is quite CPU intensive, even in an optimized algorithm, and some 9th grade algebra can save the CPU some work. In this application, it is desired to know when

Eq. (8.2) Distance = 20

which upon squaring both sides, becomes

Eq. (8.3) $\text{Distance}^2 = 400$

The $20^2 = 400$ need only be computed once, outside the loop (or by the programmer, if it is non-variable). This eliminates computation of the square root inside the loop, and is demonstrated in **Listing 8.10**.

Listing 8.10 Simplified Radial Distribution Function Calculation

```
Within_Distance = 0;
for(Particle_Counter=1;
Particle_Counter<=Number_of_Particles;    ++Particle_Counter){


      // compute distance between current
      // particle and reference particle

      tempX = X_reference - X(Particle_Counter);
      tempX *= tempX;

      tempY = Y_reference - Y(Particle_Counter);
      tempY *= tempY;

      // this is Distance²
      Distance2 = tempX + tempY;

      ' compares Distance² to 20²
      If (Distance2 <= 400) {
            ++Within_Distance;
      }

}
```

This optimization can be used in numerous places in virtually *any* scientific application. If fact, this leads to a further improvement on the Lennard-Jones pair energy.

In the algorithm for the Lennard–Jones pair energy presented in Section 8.2, r is computed using an `sqrt` function, and then r^6 and r^{12} are computed. Again, this 'back and forth' approach wastes CPU cycles. Optimization will follow the general idea that optimized code is typically more obscure (harder to read source code, at least when compared to equations written on paper).

As shown in Section 8.2, 'self multiplication' six times is an optimization. If $R = r^2$, that is, the `sqrt` is *not* taken in the distance calculation, the RISC style computation of r^6 and r^{12} becomes

Eq. (8.3) $$r6 = r^6 = r^2 * r^2 * r^2 = R * R * R$$

and

Eq. (8.4) $$r12 = r^{12} = r^6 * r^6 = r6 * r6$$

The RISC optimizations are not lost; rather three multiplications are eliminated! The pair energy calculation is thus optimized to **Listing 8.11.**

Listing 8.11 Combined r and LJ Energy calculation with no sqr() function

```
TempX  = x1  -  x2;
TempX  *=  TempX;
TempY  = y1  -  y2;
TempY  *=  TempY;
TempY  +=  TempX;

'don't need to compute sqr(r) then re-square
'it in the energy calculation.  r^2 is used
'to compute r^6 directly as r^6=(r^2)^3

r6  = TempY * TempY * TempY;
r12  = r6 * r6;

Energy = 1 / r12 - 1 / r6;
```

Results (Demo *RISC05*) comparing the combined r and LJ Energy with and without the squaring function are shown in Figure 8.3 and Table 8.2. These data demonstrate an important General Statement:

> *Whenever possible, eliminate high order functions, such as square roots, from the algorithm.*

This simple technique of 'space transformation' can be used to eliminate floating- point operations such as exp, ln, sqr, sin, cos, etc in a broad variety of modeling and curve fitting programs. This is analogous to using log or log-log graph paper to eliminate the computations when graphing by hand. The key is to focus on ways to *simplify* algorithms. At least, the needed math can often be performed *outside* large iteration loops.

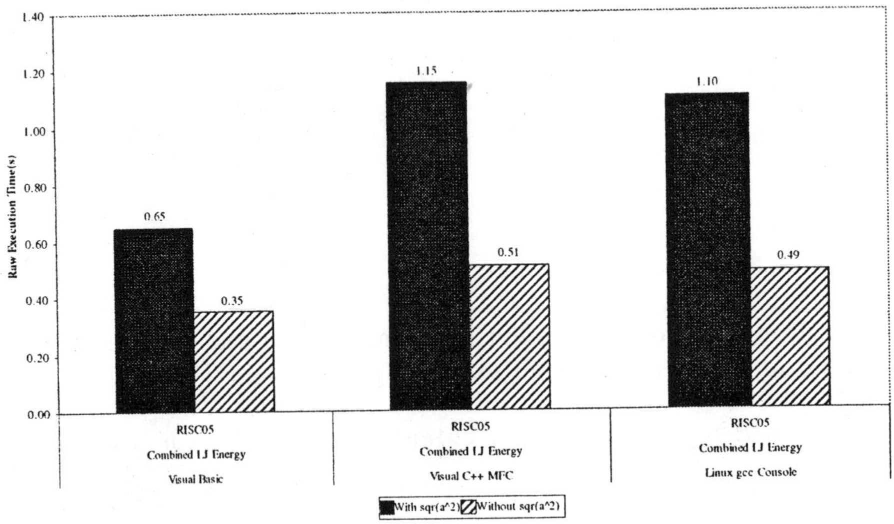

Figure 8.3: Demo *RISC05* raw execution times on the test computer for 10,000,000 iterations of the reduced Lennard-Jones energy pair energy calculation. Compared are the computations with squaring the distance and using the distance squared (without a square root in the distance step).

Table 8.2: Reduced empirical clock cycles required per iteration for the RISC calculation of the Lennard Jones energy with and without the unnecessary squaring of a square root.

	With sqr(a^2)	Without sqr(a^2)
RISC05	133	56

An example of this last point can be shown again using the Lennard-Jones pair energy. The actual equation for the total energy (for *N* particles) is

Eq. (8.5)
$$LJ\_Energy = \sum_{i=1}^{N} \sum_{j=1,\neq i}^{N} C\left[\left(\frac{\sigma}{r_{ij}}\right)^{12} - \left(\frac{\sigma}{r_{ij}}\right)^{6}\right]$$

where C is related to the strength of the interaction and s is related to the particle separation of maximum interaction. Two simplifications to this equation apply, beyond those already investigated. First, the value *C* need not appear in the loop at all since it can be factored outside the double sum; since multiplication by *C* is a floating point operation (in general), this removes $N^2 - N$ floating point muliplies *per iteration*.

The issue of using σ^6 and σ^{12} in the loop is not so simply addressed, since they cannot be fully factored. Actually, a σ^6 could be factored, but that still leaves σ^2 inside the loop. Could this be computed for each term in the double sum? Not if fast code is desired. Computing the σ^2 inside the loop is clearly undesirable.

Chapter 9: Look-Up Tables

Until fairly recently, physical memory was generally barely adequate in personal computers. As a result, most computers had memory spaces that limited the size of computations that could be performed in memory. In an interesting extreme example, one of the author's first programming experiences in the early 1980's was on a device that had 64 *bytes* of memory, and that was for both code and data! Programming styles developed during this period of 'small memory' that utilized redundant computation; it was 'cheaper' (and necessary) to re-compute information than to store it.

The evolution of PC's during this time serves as an example of how the predominate programming style evolves differently from hardware. While PC's running MS-DOS had a 640 kB 'upper limit' for physical memory (ignoring Expanded Memory and Extended Memory developments), CPU's evolved from initial 4.77 MHz clocks through about the 50-60 MHz 486 era. Scientist Programmers, especially those learning to program during this time, engrained a habit of 'saving memory.' On a system with limited data storage, size optimization was more critical than speed optimization. For example, BASIC compilers for MS-DOS code had a 64 kB limit for arrays, which is not very large for single or double precision floating point data. On such a system, data was not stored but recomputed as needed. And who really cared anyway? The CPU's were churning away at 10-50 MHz, so compute cycles were cheap.

Around the time that PC processors were hitting 50 MHz, Windows and protected mode addressing became popular, and the 640 kB barrier was broken. At the time of this writing, the situation is much different from the 'fast CPU, low memory' system popular in the early 1990's. Though PC CPU's cost/FlOPS have also decreased, memory cost/MB has decreased to the point that PC's typically have 128 MB, 256 MB or larger physical memory spaces. That is, during the time that CPU clocks increased by a factor of 50 -100, "normal" memory spaces increased by a factor 100-500. Code written in the 1980's or early 1990's style is not optimized for modern architectures – especially the large available physical address space. Further, programming new codes in this style does not use the available hardware to best advantage.

In Chapter 9, the Look-Up Table (LUT) technique is explored. A LUT is essentially a very large data structure comprised of pre-computed arrays or matrices. This technique is extremely memory intensive, but yields significant performance gains.

Coupled with the RISC techniques in Chapter 8, The LUT style program should run roughly 1000 times faster (or more) than codes without these techniques employed. The fundamental essence of the LUT is that memory access, especially when the L1 cache is employed, is generally faster than floating point computation.

9.1 MEMORY ACCESS VS. COMPUTATION

Regardless of specific CPU architecture, a 'fetch from memory' instruction executes faster than a general floating-point compute instruction. Table 9.1 shows some 80287, 486 and Pentium FPU instructions and latencies. Of particular note is the comparison

Table 9.1 Execution latencies for several x87 computational and memory access instructions for the 286, 486 and Pentium processors. Recall that for many x87 instructions, the Top of Stack register is implied.

Instruction Mnemonic	Operation	287 Latency	486 Latency	Pentium Latency
F2XM1	$2^x +1$	310-630	140-279	13-57
FADD mem64	memory add	95-125	8-20	3/1
FCHS	change sign	10-17	6	1
FCOS	cos(x), $0<x<\pi/4$	--	257-354	18-124
FDIV mem64	memory division	221-231	73	39
FLD mem64	memory to register	40-60	3	1
FLDZ	push 0.0 onto register stack	11-17	4	2
FLD1	push 1.0 onto register stack	15-21	4	2
FLDL2E	push $\log_2 e$ onto register stack	15-21	8	5/3
FLDL2T	push $\log_2 10$ onto register stack	16-22	8	5/3
FLDLG2	push $\log_{10} 2$ onto register stack	18-24	8	5/3
FLDLN2	push $\log_e 2$ onto register stack	17-23	8	5/3
FLDPI	push π onto register stack	16-22	8	5/3
FMUL mem64	memory multiply	154-168	14	3/1
FPATAN	partial arctangent(x), $0<x<\pi/4$	250-800	218-303	17-173
FPREM1	IEEE compatible remainder	--	72-167	20-70
FPTAN	partial tangent(x), $0<x<\pi/4$	30-540	200-273	17-173
FSIN	sin(x), $0<x<\pi/4$	--	257-354	16-126
FSQRT	square root	180-186	83-87	70
FSUB mem64	memory subtraction	95-125	8-20	3/1
FYL2X	compute $Y^*\log_2 X$	900-1100	196-329	22-111

of the `FLD` mem64 instruction to those performing numerical operations, such as `FSIN`. In this case, the `FSIN` instruction is 16-126 times *slower* than an `FLD`. Clearly, the processors have evolved so the 'gap' between a single compute and fetch has narrowed, but the gap remains nonetheless. Further, it should be remembered that, in general, a real computation involves numerous floating-point operations. If a desired result can be obtained via a single memory fetch rather than computation, the code executes faster. How is this implemented in code?

The data must be computed at some point in the job, but the key is to compute it *once* and store the values in physical memory. For a given application, this may mean pre-computing values that never get used, but the CPU overhead for generating the LUT is tiny compared to the overall job. The data stored in memory is used repeatedly throughout subsequent steps of the calculation; this is the Look-Up Table.

9.2 THE LUT IN ACTION – 1 DIMENSIONAL EXAMPLES

Though the use of Look-Up Tables is straightforward, it is instructive to present several practical examples to illustrate the coding style. In this section, the Lennard Jones energy is presented as a detailed example. In addition, several common functions are illustrated and the error introduced when using this technique is briefly summarized.

9.2.1 Lennard Jones Energy

The iterative Lennard Jones energy computation, as used in a Monte Carlo procedure, is a good candidate for use of a LUT. For example, in a Monte Carlo calculation involving 10,000 configurations of a single 'box' of 1000 particles, roughly 50,000,000,000 individual pair energies must be computed if the total energy is explicitly computed for each configuration. Each of these individual pair energy computations involves numerous floating-point operations (see Chapter 8).

To eliminate these billions of energy calculations (many of which are redundant, actually) 'all possible' individual pair energies are computed and stored in an array indexed on particle separation. Here, 'all possible' refers to the spatial domain of the stated problem at a practical precision. For the present example, INDEX = r_{ij} * 10 can be used for a mesh of 0.1 pm if r is expressed in pm. For a box that is 2000 pm x 2000 pm, the largest distance two particles can be separated is 2000 * sqr(2) = 2828 pm, so the array will need 28,290 entries. In x86 based computers, each double precision floating-point value requires 8 bytes (64 bits), so the array data (not counting the array overhead) requires 226,320 bytes.

Execution times for Demo *LookUp01* that implements the Lennard Jones Energy LUT are shown in Figure 9.1. The optimized RISC algorithm from Chapter 8 was applied to the array initialization. The array is only initialized once per execution. In addition, the cycles used to compute the array are not 'wasted' since these

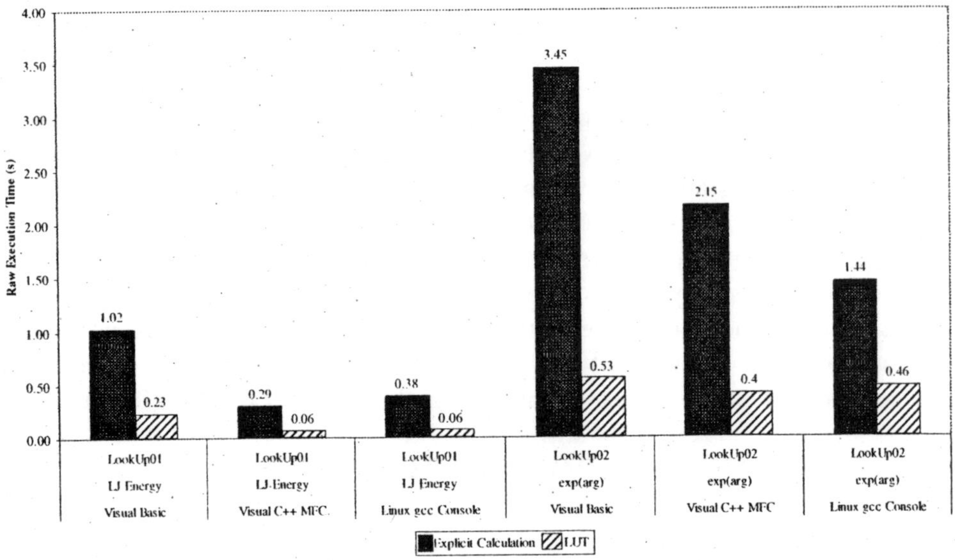

Figure 9.1: Raw execution times on the test computer for 10,000,000 iterations computation vs. look-up of reduced function values. Compared are the Lennard Jones pair energy and Boltzmann factor functions, as computed in Demos *LookUp01* and *LookUp02*.

computations (or at least most of them) would eventually need to be done anyway. The point is that this is the maximum number of energy computations that need to be performed; additional energies can be obtained from the array. The two code structures demonstrated in Figure 9.1 are **Listing 9.1** and **Listing 9.2** each wrapped by a loop counter and using a constant for *r*.

Listing 9.1 Explicit Calculation of pair energy each time through the loop

```
// r is selected in code

// Compute Energy using RISC style from
// Chapter 8

r6 = r * r * r * r * r * r;
r12 = r6 * r6;

ThisEnergy = 1 / r12 - 1 / r6;
```

Listing 9.2 Using look-up table, precomputed and stored in array En(r)

```
// the array En[r] is precomputed
```

```
// for r=0 to 2829.0 pm
// r is selected in code; r is a double

Index = (int)(r * 10);

ThisEnergy = En[Index];
```

Given the very clear performance gain, 226 kB of RAM is a cheap trade-off to eliminate millions of floating point operations in this simple example. Note that even this small LUT may not have been possible on MS-Dos machines running about ten years ago, depending upon the compiler.

This example further illustrates the obscurity idea repeatedly mentioned throughout this book regarding high performance code. A casual glance at the source code in **Listing 9.1** reveals what *ThisEnergy* is: $1/r^{12} - 1/r^{6}$. However, a similar glance at **Listing 9.2** may leave the reader bewildered regarding what *En* is. It is clear that *ThisEnergy* is simply an array entry from *En*(*r*), but what is stored in that array? Good programming practices such as systematic variable naming and clear comments in the source code limit the obscurity cost to some degree.

9.2.2 Other 1-D examples

The LUT can be applied to any situation provided enough physical memory exists to store the data. For example, in the Monte Carlo procedure, the Boltzmann factor $e^{-energy/kT}$ is repeatedly computed; why not precompute and store these as well? Other examples include gaussian distribution functions, error functions or any function that involves complicated floating point arithmetic. *LookUp02* demonstrates the Boltzmann factor Look-Up Table with results shown in Figure 9.1. Look-Up Tables for several common floating-point functions are demonstrated in *LookUp03*, and execution times are summarized in Figure 9.2.

9.2.3 Introduced Error in the LUT Technique

By using energies only at discrete values, some error is introduced (unless the look-up contains an entry for each value in the precision used). For example, consider the energy LUT defined at 0.1 angstrom intervals, r_{LU} = 0.1, 0.2, 0.3 ... *MaxR*. If the computational problem generates a value r_{actual} = 2.36 angstroms, precision is lost: this value will enter into the LUT as either 2.3 or 2.4, depending upon how the code is written. The error is greatest where the first derivative of the LUT function is large; for the Lennard Jones pair energy function, the error is greatest for values of r near σ. A consideration of the error introduced should be carefully given for any computational problem for which the LUT technique is being contemplated.

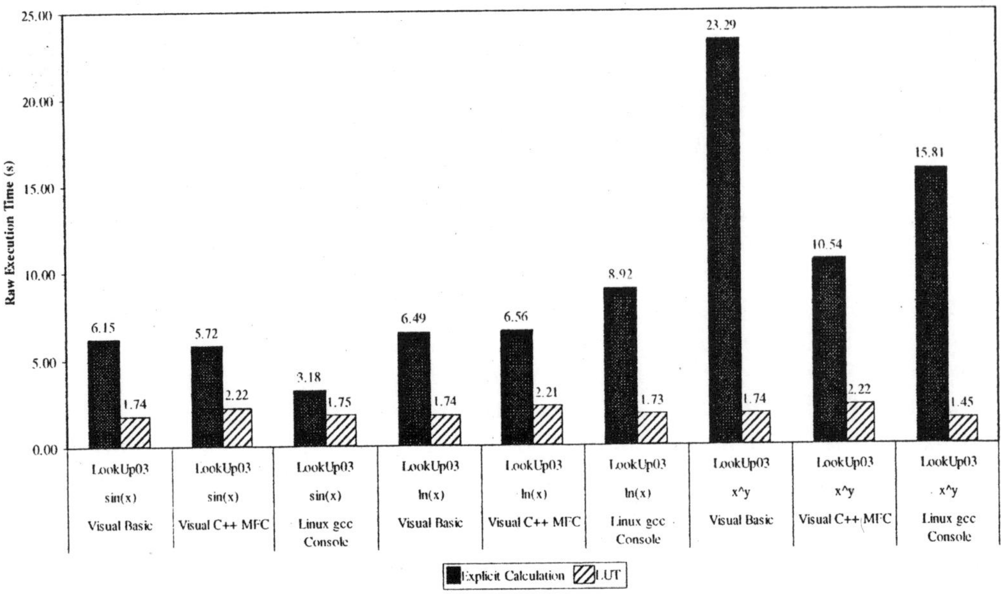

Figure 9.2: Demo *LookUp03* raw execution times on the test computer for 50,000,000 iterations computation vs. look-up of several common floating point functions.

9.3 THE LUT IN ACTION - 2-DIMENSIONAL LOOK-UPS

In the previous section, the use of an array to store function values rather than computing them was outlined and demonstrated. The idea can easily be extended to multi-dimensional functions or functions of more than one variable. Two specific examples are given: the distance, r(x,y) and the heterogeneous Lennard Jones pair energy, $E(C, \sigma, r)$.

9.3.1 Distance

It might be tempting to code the Total Energy calculation in the developing Monte Carlo procedure as **Listing 9.3** for N particles:

Listing 9.3 **Combined r and LJ Energy using LJ Energy Look-Up Table**

```
// En(r) is array initialized to
// 1/r^12 - 1/r^6

// a value for the index
int rIndex;
```

```
// some made-up values for the
// two coordinates
double x1=1.234;
double x2=2.675;
double y1=0.045;
double y2=2.073;

// temps for distance calc
double TempX;
double TempY;

TempX = x1 - x2;
TempX *= TempX
TempY = y1 - y2;
TempY *= TempY;
TempY += TempX;

r = sqrt(TempY);
rIndex = (int)r * 100;

ThisEnergy  = En[rIndex];
```

This certainly improves the algorithm by looking-up the energy at each step rather than using explicit computation. However, there remains significant floating-point effort expended on the calculation of the distance *r*. Having opened the Look-Up Table door, there is little reason to include the distance calculation in the loop. An array holding distances in the LUT should improve the overall performance even further. To implement this, a two dimensional array is needed so that the indices are Δx and Δy. This array allows the Look-Up of distance. To save both physical memory and compute cycles, the distance Look-Up array is of integer type since what is actually being looked-up is the index with which to enter the *Energy* array. Therefore, an error must be considered, and the array step must be carefully planned so that a significant loss of precision does not occur. Consider, for example, **Listing 9.4** and **Listing 9.5**.

Listing 9.4 Initialize LUT for r as two dimensional array

```
// The array itself is RVal and is
// declared as an int so type conversions
// are not done inside the iterative loop
// (that is, when looking up Energy(RVal))

for(deltaXCounter=1; deltaXCounter<=2829;++deltaXCounter){

      TempX = (double)deltaXCounter;
```

```
        TempX  *= TempX;

        for(deltaYCounter=1; deltaYCounter<=2829;
            ++deltaYCounter){

            TempY  =  (double)deltaYCounter;
            TempY  *=  TempY;

            RTemp  =  sqrt(TempX + TempY);

            Rval[deltaXCounter][deltaYCounter]
                 =  (int)Rtemp;

        }

}
```

Listing 9.5 **Actual LJ Energy Function using LUT for both r and Energy**

```
double Homog_LJ_Energy(){

        // Declarations go here
        // En and Rval are arrays declared
        // and initialized outside this
        // function.

        ThisEnergy = 0;

        // N is number of particles

        for(I=1; I<=N-1; ++I){

            for(J=I+1; J<=N; ++J){

                deltaX = (int)(fabs(X[I] - X[J]);
                deltaY = (int)(fabs(Y[I] - Y[J]);

                R = Rval[deltaX][deltaY];

                ThisEnergy += Energy[R];
            }

        }

        return(ThisEnergy);
}
```

This sample code uses a 1 pm mesh size in contrast to the 0.1 pm in **Listing 9.2**. With a 1 pm mesh, the distance Look-Up Table requires 32 MB, whereas the 0.1 pm mesh would require 3200 MB! This is a definite scalability issue; the LUT scales as (significant figures)^2. With a 1 pm mesh for the distances, there is no need to store the Energy Look-Up Table at 0.1 pm precision, so the Energy array in **Listing 9.5** can be defined using a 1 pm mesh as well.

The function "Homog_LJ_Energy" as written in **Listing 9.5** contains *no explicit complicated floating - point math!* Though there are floating point adds, the latency tables show FADD can execute in as little as one cycle on Pentium class processors. If this function is wrapped by an iterative procedure, the Lennard Jones energy can be "computed" millions of times without using CPU cycles redundantly computing the pair energies. That is, in this example, there are only 282,900 *possible* individual pair energies and $2829^2/2$ = 4,001,621 *possible* separation distances (with its 640 kB limitation, this was clearly impossible under MS-Dos!). Though the computation will certainly involve more energies in the total procedure, the additional ones are additive combinations of these 'basis' energies. Without the LUT, the energies and distances must be computed millions of times each; with the LUT, energies and distances are computed once. Since $RVal_{ij} = RVal_{ji}$, the $RVal$ matrix is symmetric and only half the data really need be stored. The trade-off for this is that some logic would then need to be included to 'swap' the values of x and y as needed. Figure 9.3 shows raw execution times comparing the RISC portion of **Listing 8.11** to **Listing 9.5** as computed using Demo *LookUp04*.

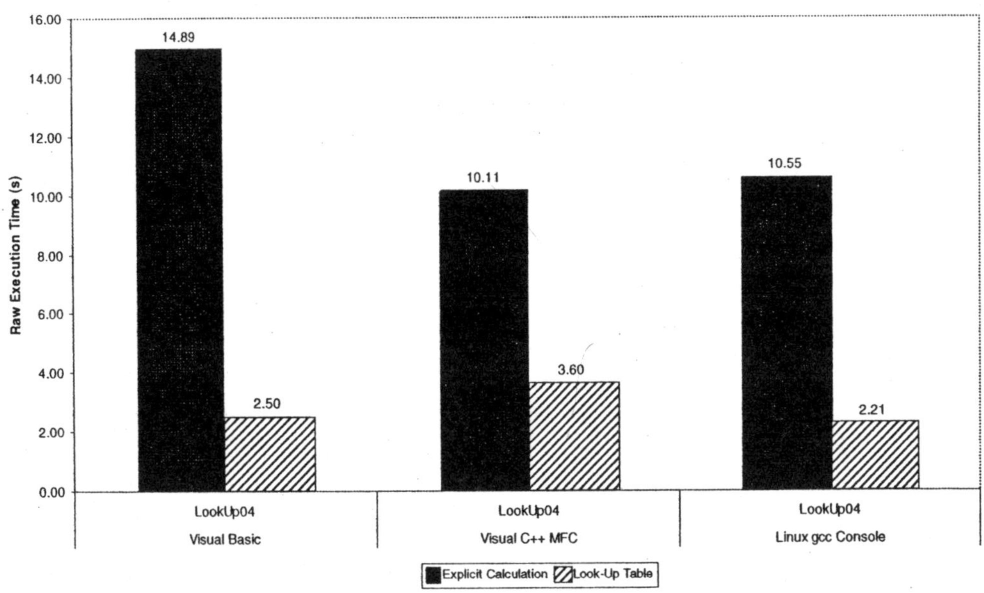

Figure 9.3 Demo *LookUp04* raw execution times on the test computer for 100,000,000 iterations for the combined distance (r) and Lennard Jones energy computation vs. look-up.

9.3.2 Heterogeneous Lennard Jones Energy

To this point in the Monte Carlo project development, only homogeneous or 'monoatomic' systems have been modeled. The Look-Up Table approach enables a convenient economical solution to extending the energy calculation to microscopically heterogeneous systems. The simplest such system, that consisting of two types of particles, requires three types of interactions:

Type 1	A — A
Type 2	B — B
Type 3	A — B

The energy array entry to the LUT for this system is a 3 x [BoxSide * sqr(2)] matrix, so the memory overhead is only three times larger. To extract the energy from this LUT, an interaction array can be used, as depicted in **Listing 9.6**.

Listing 9.6 Initialization of Heterogeneous LJ Energy Look Up Table

```
// storing energy table with multiple
// interactions declarations go here

// Depth(a) is array holding three interaction
// 'strengths'
// Sigma(a) is array holding three interactions
// 'separations'

// initialize the sigma6 and sigma12
// arrays outide the loop
Sigma6[1]  = Sigma[1] * Sigma[1] * Sigma[1] *
             Sigma[1] * Sigma[1] * Sigma[1];
Sigma12[1]  = Sigma6[1] * Sigma6[1];

Sigma6[2]  = Sigma[2] * Sigma[2] * Sigma[2] *
           Sigma[2] * Sigma[2] * Sigma[2];
Sigma12[2]  = Sigma6[2] * Sigma6[2];

Sigma6[3]  = Sigma[3] * Sigma[3] * Sigma[3] *
           Sigma[3] * Sigma[3] * Sigma[3];
Sigma12[3]  = Sigma6[3] * Sigma6[3];


for(Rcounter=1; Rcounter<=MAXR; ++Rcounter){

     R = (double)(RCounter/100);
```

```
    R6  =  R  *  R  *  R  *  R  *  R  *  R;
    R12  =  R6  *  R6;


    E[1][Rcounter]  =  Depth[1]  *  (Sigma12[1]/R12  -
          Sigma6[1]/R6);


    E[2][Rcounter]  =  Depth[2]  *  [Sigma12[2]/R12  -
          Sigma6[2]/R6];

    E[3][Rcounter]  =  Depth[3]  *  (Sigma12[3]/R12  -
          Sigma6[3]/R6);

}
```

Implementing the heterogeneous case will be explored in further detail in Chapter 10 Section B – Type Flag Encoding.

9.4 VIRTUAL VERSUS PHYSICAL MEMORY

In order for the Look-Up Table technique to yield an actual performance increase, the Look-Up Table itself must generally fit into the available physical memory. For example, in Section 9.3.1, it was noted that that the distance matrix for a 2000 x 2000 pm box with a 0.1 pm mesh would require 3200 MB of memory for double precision data. This exceeds the amount of physical memory on many personal computers, but may lie within the virtual memory space of many. However, a computation that repeatedly calls data from such a large 'swap file' will be quite slow.

Modern hard drive equipment and Operating Systems display considerable performance over older drives partly due to caching; often-needed data is stored in physical memory. However, the premise with the Look-Up Table is that the data is *randomly* selected from the table. Therefore, caching the data is not very effective, and some data will be swapped in and out repeatedly. In this situation, the system is working hard pushing the same data along the Front Side Bus between physical memory, CPU and disk.

In the case where the physical memory requirement exceeds the amount of physical memory available, it is generally better to actually perform the floating-point computations than utilize a Look-Up Table. This can be decided at run time via a call to the API function (for Windows) `GlobalMemoryStatus`, which returns considerable data about the memory system into a `MEMORYSTATUS` structure. The contents of this structure are summarized in Table 9.2. In Linux, one way to access available physical memory is to read /proc/meminfo. One then would subtract the `Active` value from the `MemTotal` value. In summary, if the desired Look-Up Table

Table 9.2: Members of the MEMORYSTATUS structure into which the GlobalMemoryStatus Windows API function returns data. A pointer to the structure is passed in the function call.

Member	Brief Description
DwLength	Size, in bytes, of the `MEMORYSTATUS` structure
DwMemoryLoad	Between 0 and 100, an estimate of percentage of current memory in use
DwTotalPhys	Total number of bytes of Physical Memory in the system
dwAvailPhys	Number of bytes of Physical Memory available
dwTotalPageFile	Number of bytes that can be stored in the paging file (not the size of the page file on disk)
dwAvailPageFile	Number of bytes available in the paging file
dwTotalVirtual	Total number of bytes in the user mode portion of the virtual address space of the *calling process*
dwAvailVirtual	Number of bytes of uncommitted memory in the user mode portion of the virtual address space of the *calling process*

is smaller than the available physical memory (the `dwTotalPhys` member in Windows), the Look-Up Table will be faster. If not, computation may be faster.

Chapter 10: Other Algorithm Optimization Techniques

In the previous Part II chapters, specific optimizations were considered. Numerous optimizations are not easily categorized in one of the headings of Chapters 5-9. These might be collectively termed algorithmic optimizations, as they constitute fundamental changes to the algorithm that are not simply implementation techniques. The field of designing and improving actual algorithms is beyond the level of effort most scientists wish to employ in their code, and in general, this is beyond the scope of this book. However, several techniques are included here to demonstrate the general idea, especially in terms of optimization as an evolutionary process.

10.1 USE OF SYMMETRY

The use of problem symmetry was introduced in Section 1.4. For example, in a pair energy calculation, the energies are symmetric and the 'naïve' approach to computing them in the Scientist Style wastes CPU effort by computing twice. The size of the loops used, and hence the number of computations to be performed, can be made smaller by applying this symmetry. There are many other examples of compute-space symmetry that can be applied.

As another simple example, consider the electric scalar potential field for a cylindrical electro-optic lens. A sample lens, with representations showing reduction of the problem by symmetry, is shown in Figure 10.1. The full 3-dimensional problem is illustrated in Figure 10.1(a). The scalar potential *can* be solved for all points in the 3-d domain, but this has two significant disadvantages. First, the absolute number of points is relatively large for a given mesh representing the domain. Second, the 3-d mesh solution scales roughly as $m \times n \times p$, for the rectangular solid of length m, width n and height p that contains the cylindrical lens. As the mesh is refined, or the domain gets larger, the computational effort increases dramatically.

Figure 10.1(b) shows reduction of the problem to two dimensions by using the axial symmetry of the problem. The potential is independent of the angle around the

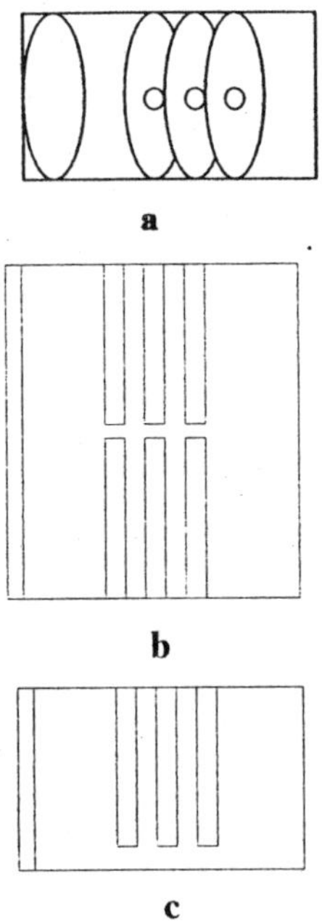

Figure 10.1: Cylindrical electrooptic lens used in Demo *Other01*. 10.1(a) is the complete lens system. The lens system reduced by the cylindrical symmetry, but with both axial reflections, is shown in Figure 10.2(b). Figure 10.1(c) shows only the symmetry unique portion of the problem.

cylinder axis, so computation of the potential along concentric circles is wasted effort. This significant reduction of the problem domain improves the scalability, which is reduced from m x n x p to m x p. A further improvement results in reducing the factor p to $p/2$ by again using the axial symmetry of the cylinder. This is shown in Figure 10.1(c).

This application is demonstrated in Demo *Other01* for the simply einzel lens depicted in Figure 10.1. In *Other01*, the solution domain is as shown in Figure 10.1(c), but the graphical representation is shown as Figure 10.1(b). This is accomplished simply by reflecting the solution within the drawing function. The demo uses the Point Gauss-Seidel method to iteratively solve Laplace's Equation for the electric potential.

The only danger to avoid is arbitrarily applying symmetry when one 'thinks' it should be present. For example, one test of a theoretical model in quantum chemistry calculations is to *not* impose symmetry on the computation to determine if the simulation correctly predicts the known molecular symmetry (as evidenced by experiment). Imposing symmetry restrictions in *any* calculation does reduce the compute time, but may introduce artificial errors.

10.2 ELIMINATION OF NESTED LOOPS

The Monte Carlo process used as a case study throughout this book involves computation of the Lennard Jones energy each time a random particle is moved to a

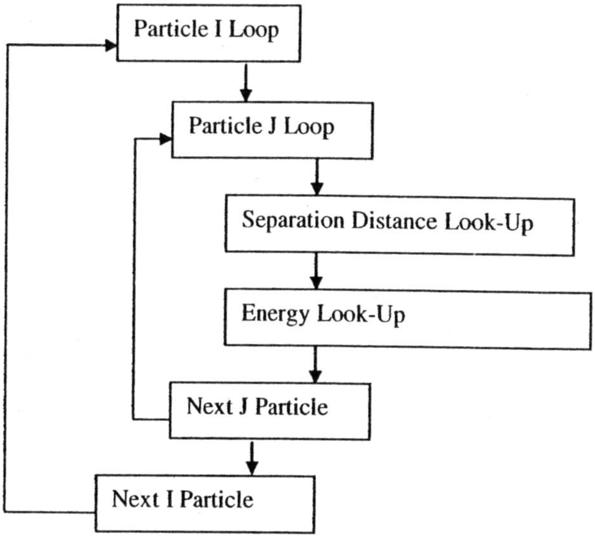

Figure 10.2: Flow chart summarizing the total system energy (as a sum of Lennard Jones pair energies) by an algorithm using a nested loop structure. This computation scales as N^2 for N particles.

random new position. As described thus far, the full energy is obtained each iteration inside a nested loop structure. That is, *after* the random particle is moved, the energy is computed as shown in Figure 10.2. As mentioned in Chapter 1, nested loop calculations do not scale favorably as the number of particles increases. A means to improve the scalability is now presented.

The Lennard Jones energy at each step can be written

Eq. (10.1) $$Energy = Moved\_Particle\_Part + Static\_Part$$

where Moved_Particle_Part is

Eq. (10.2) $$Moved\_Particle\_Part = \sum_{i=1, i \neq j}^{N} \left[C_n \left(\left(\frac{\sigma_n}{r_{ij}} \right)^{12} - \left(\frac{\sigma_n}{r_{ij}} \right)^{6} \right) \right]$$

for N particles, n interactions between particle i and j and j is the moved particle. The static part of the energy is therefore

Eq. (10.3) $$Static\_Part = \sum_{i=1, i \neq j}^{N-1} \sum_{k=i+1, k \neq j}^{N} \left[C_n \left(\left(\frac{\sigma_n}{r_{ik}} \right)^{12} - \left(\frac{\sigma_n}{r_{ik}} \right)^{6} \right) \right]$$

The important point to note is that the Static_Part is the same as before the particle was moved. Since this "static part" has not changed by moving one particle, there is

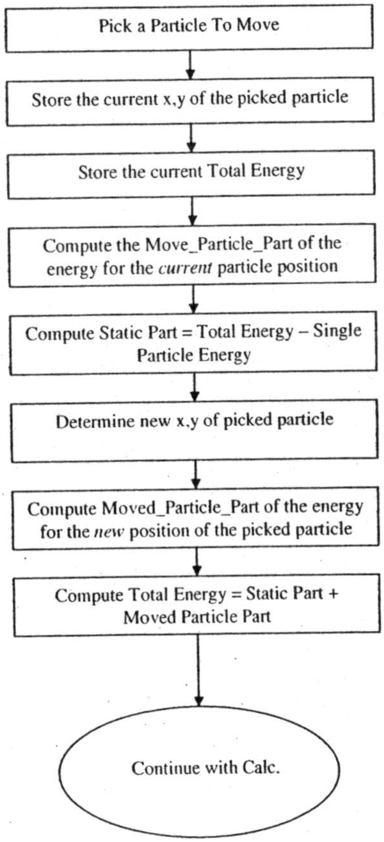

Figure 10.3: Flow chart summarizing the total system energy (as a sum of Lennard Jones pair energies) by an algorithm that does not use a nested loop structure. This computation scales as 2N for N particles.

no reason to compute it each time a new particle is selected. To take advantage of this optimization, the algorithm flow must be rewritten, as shown in Figure 10.3. For this algorithm, the `Single_Particle_LJ` function shown in **Listing 10.1** is used:

Listing 10.1 Computing only energy change in Single Particle LJ Energy routine

```
double SingleParticleLJ(int MovedIndex, int
     TotalNum, int Xmoved, in Ymoved){

     // Declarations here
     // particles is a structure
     // containing the positions, x and y

     ThisEnergy = 0;

     for(ParticleCounter=1; ParticleCounter<=N;
```

```
        ++ParticleCounter){

        rx=particles[ParticleCounter].x;
        ry=particles[ParticleCounter].y;

        If  (ParticleCounter  !=  MovedIndex){

            deltaX  =  abs((int)(rx  -  XMoved));
            deltaY  =  abs((int)(ry  -  YMoved));
            R  =  Rval[deltaX][deltaY];
            ThisEnergy  +=  Energy[R];

        }

    }

    return(ThisEnergy);

}
```

The code associated with the above algorithm in Figure 10.3 (shown in **Listing 10.1**) bears little resemblance to the pure Scientist Programmer style shown in **Listing 1.1** . More importantly, however, the scalability of this algorithm is much improved from N^2 to a time roughly proportional to 2N (at least for the Lennard Jones portion of the total procedure). Demo *Other02* compares the algorithms that scale as N^2 and 2N, and raw execution data versus number of particles is shown in Figure 10.4.

10.3 REDUCING DECISION LOGIC OVERHEAD - BIT FLAG ENCODING

One of the most important logical tasks a computer can perform is decision making.

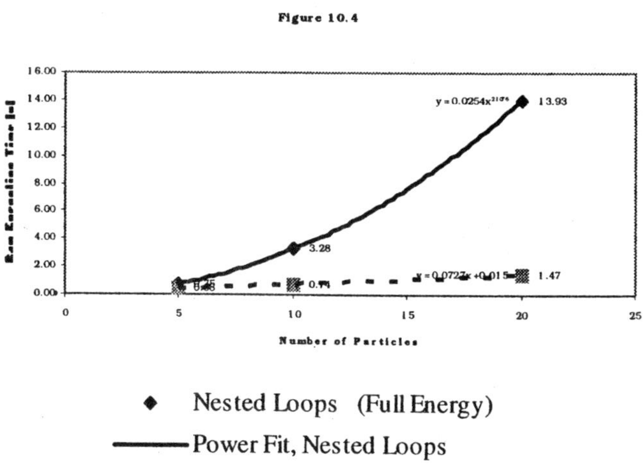

Figure 10.4: Demo *Other02* raw execution times for 50,000 iterations of the total energy computations depicted in Figure 10.2 and Figure 10.3.

If the capability to branch to different program procedures based on system state did not exist, many useful programs would not be possible. However, decision making/ branching can significantly reduce performance. For even a simple decision/branch, the CPU must compare two terms (which is often done via subtraction or a bitwise logical operation) and jump to the appropriate block of code based on the outcome. Jumping, either conditional or unconditional, is discussed briefly in Chapter 2, and was shown there to have the potential for being a high latency operation. Further, CPU features such as bus pipelines lose their effectiveness when a `JMP` (or related) instruction is encountered. This issue is of such importance that CPU designers have included advanced circuitry, such as Branch Target Buffers and associated branch prediction algorithms, in modern PC CPU's to compensate for the performance loss. Even with the advanced hardware designs, programs without extensive decision/ branching code will execute faster than programs with a large number of decisions/ branches.

Some real computational codes currently in distribution contain extensive decision/ branching blocks of code. For example, one *ab initio* Quantum Chemistry program contains 19 `if` blocks in a row in only one (of many) source file. To develop a technique to reduce this decision/branching overhead, the heterogeneous Lennard Jones pair energy Look-Up Table introduced in Chapter 9 is used. For two particle types, A and B, one way to implement the heterogeneous LUT entry for the energy is to define a data structure for the particles. Such a structure for two particle types can be declared as shown in **Listing 10.2**:

Listing 10.2 Sample User Defined Data Type for Heterogeneous (multi-particle) LJ Energy

```
// This goes in declaration section
struct PARTICLE
{
      bool Type;
      double X[2829];
      double Y[2829];
}

PARTICLE Particles[NumberOfParticles];
```

To use this, *Particles[i].Type = True* if *i* is an A particle and is *False* if *i* is a B particle. Using this structure to map into an energy interaction type (the first dimension in an *Energy[Type][r]*) array requires some logic. One Scientist Style approach, still assuming two particle types, is shown in **Listing 10.3**:

Listing 10.3 Heterogeneous LJ Energy via If...Then structure to select interaction type

```
// Note: Test Expression is evaluated multiple
// times
// access the Ith and Jth particle types via
// Particles(I or J).Type

if (Particles[I].Type) &&
        (Particles[J].Type) EnergyType = 1;

if (!Particles[I].Type) &&
        (!Particles(J).Type) EnergyType = 2;

if (Particles[I].Type) &&
        (!Particles[J].Type) EnergyType = 3;

if (!Particles[I].Type) &&
        (Particles(J).Type) EnergyType = 3;
```

This approach locks the code into performing many compares and branches, but does have the Scientist Programmer style advantage of being relatively readable. The CPU overhead for the logical testing itself can be costly, especially if the 'Particle Type-Space' is greater than 2. Table 10.1 shows the number of EnergyTypes (Energy Type Space) for several small Particle Type Spaces (that is, the number of types of particles). The Energy Type Space is equal to the *minimum* number of possible 'true' branches in an if control structure, and the reader can verify that the complexity of such a control structure increases quite rapidly with increasing number of particle types. Also, note that although the example in **Listing 10.3** has four comparisons, not three (there are three energy types), since there are two ways *EnergyType* = 3 can be constructed. Therefore, the number of branches required grows *much* more rapidly than the Energy Type Space!

A higher performance approach can be used to eliminate the logical testing from the procedure completely. By changing the Particle.Type variable type from Boolean to Integer, the particle type can be encoded as a numerical value: 0 for Type A or 1 for Type B. Though a style similar to **Listing 10.3** could still be used (testing for numerical values rather than true/false), a better approach is to compute EnergyType from

Eq. (10.4) EnergyType = Particles[I].Type + Particles[J].Type

which yields EnergyType = 0 for an A—A interaction, 1 for an A—B interaction and 2 for a B—B interaction. With this approach, the value computed for EnergyType can enter as the Energy array index directly, without further manipulation. This has numerous advantages over the style shown in **Listing 10.3**. First, *all* decision structures and branching is eliminated. Second, all complicated logic in **Listing 10.3** is replaced by a 'single' simple integer addition, in keeping with the idea that simpler source structures compile to faster object code.

This approach can be extended to the higher particle spaces shown in Table 10.1 with minor modification. The logic indicated by Eq. (10.4) does not work for Base-10 integer data for Particle Type. For example, suppose the Particle Type Space is 3; there are three kinds of particles in the system. Thus, there are 6 EnergyTypes possible (Table 10.1), and the simple Base-10 additive Particle Type to Energy Type mappings are shown in Table 10.2. The problem with this mapping is that it is *not*

Table 10.1: Energy Type Spaces for given Particle Type Space

Particle Type Space	Energy Type Space
1	1
2	3
3	6
4	10
5	15
n	$\left[\sum_{i=1}^{n} i\right]$

Table 10.2: Base-10 Additive Particle.Type mapping to EnergyTypes for Particle Type Space = 3. This mapping is not 1:1 since two different sets of Particle.Type map to EnergyType=2.

Particle (I). Type	Particle (J). Type	Base-10 Additive Energy Type
0	0	0
0	1	1
0	2	2
1	1	2
1	2	3
2	2	4

1:1. Two different particle type interactions map to a single EnergyType: both B—B and A—C map to EnergyType=2. This 'breaks' the algorithm as a model of a real physical system. To repair this, the idea of binary flag encoding is used. This encoding is used extensively in system programming.

Table 10.3 shows the Base-10 (decimal) and Base-2 (binary) encodings for Particles.Type for Particle Type Spaces of 3 (to compare directly to previous example) and 4 (for further clarity). In the binary 'flag' representation, each binary digit represents a true/false flag for a given particle type (1=true, 0=false):

Particle Type A	bit 0 = 1, all others 0
Particle Type B	bit 1 = 1, all others 0
Particle Type C	bit 2 = 1, all others 0
Particle Type D	bit 3 = 1, all others 0
etc	

Table 10.3: Flag Encoding Particle Types in Base-10 and Base-2 for Particle Type Spaces 3 and 4.

3 Particle Types			
	Particle Type	**Base-10 Representation (decimal, binary)**	**Base-2 Representation (decimal, binary)**
	A	0, 00000000	1, 00000001
	B	1, 00000001	2, 00000010
	C	2, 00000010	4, 00000100
4 Particle Types			
	A	0, 00000000	1, 00000001
	B	1, 00000001	2, 00000010
	C	2, 00000010	4, 00000100
	D	3, 00000011	8, 00001000

Implied in this notation is that one and only one bit is true in each Particle.Type integer. If the data management code, particularly the Particle.Type initialization, conforms to this, the EnergyType value can be easily derived in one of two ways. First,

Eq. (10.5) EnergyType = Particles[I].Type + Particle[J].Type

can be used. This mapping of Particles.Type pairings to EnergyType is 1:1 and is shown in Table 10.4. This method has the disadvantage that EnergyType must be one bit larger than the highest bit of Particles.Type, meaning at most 7 particle types can be encoded with a short integer and 15 for a long integer.

Table 10.4: Binary Particle.Type additive and OR mappings to EnergyTypes. Note that the additive mapping, while 1:1, requires an extra bit in EnergyType while wasting (not using) encoding EnergyType = 1.

Particle(I) Type	**Particle(I) Type in binary representation (decimal, binary)**	**Particle(J) Type**	**Particle(J) Type in binary representation (decimal, binary)**	**EnergyType = Particle(I) Type + Particle(J) Type**	**EnergyType = Particle(I) Type OR Particle(J) Type**
A	1, 00000001	A	1, 00000001	2, 00000010	1, 00000001
A	1, 00000001	B	2, 00000010	3, 00000011	3, 00000011
A	1, 00000001	C	4, 00000100	5, 00000101	5, 00000101
B	2, 00000010	B	2, 00000010	4, 00000100	2, 00000010
B	2, 00000010	C	2, 00000010	6, 00000110	6, 00000110
C	4, 00000100	C	4, 00000100	8, 00001000	4, 00000100

An alternative approach, which does not have this 'extra bit' limitation, is to use the logical OR bitwise operator. The 1:1 mapping in this case is different from the additive mapping and is shown in Table 10.4. This example introduces the value of bit operators for increasing performance; an arbitrary Particle Type Space can be uniquely mapped to an EnergySpace without the use of multiple, complicated logical branching. However, care should be taken when initializing the Energy array to ensure that the mapping is correct. This means the Interaction and Sigma arrays (from Chapter 9) must be carefully planned as well.

This example shows that the higher performance algorithm requires much more thought and planning on the part of the programmer. The Scientist Style tends to favor minimal data structure planning and thus does not lend itself well to bit flag encoding. Again, the source of the Scientist Style program is more 'readable' at the expense of performance. Results for a demo of this technique, *Other03*, are shown in Figure 10.5. The demo compares raw performance data for a Particle Type Space of 3 using the `if…else if…else` logic vs. the use of bit flag encoding. In BASIC, the flag-encoded version ran a little over 10 times faster; for the (non-optimized) C++ demo, the encoded version ran nearly 3 times faster.

It is important to note another feature of the bit encoded algorithm: favorable scalability with increasing particle types. From Table 10.1, it can be seen that the *minimum* number of comparisons grows rapidly for the 'traditional' if approach to the heterogeneous problem. Therefore, the run time performance of the code scales at least as bad as the number of comparisons. On the other hand, for the flag encoded algorithm, using either Equation (10.5) or the logical OR, there is *no* run time cost to increasing the particle type space. In other words, the scaling factor is a constant (there is no N dependence). This rare scaling property makes bit flag encoding an attractive approach for *any* algorithm to which it can be applied.

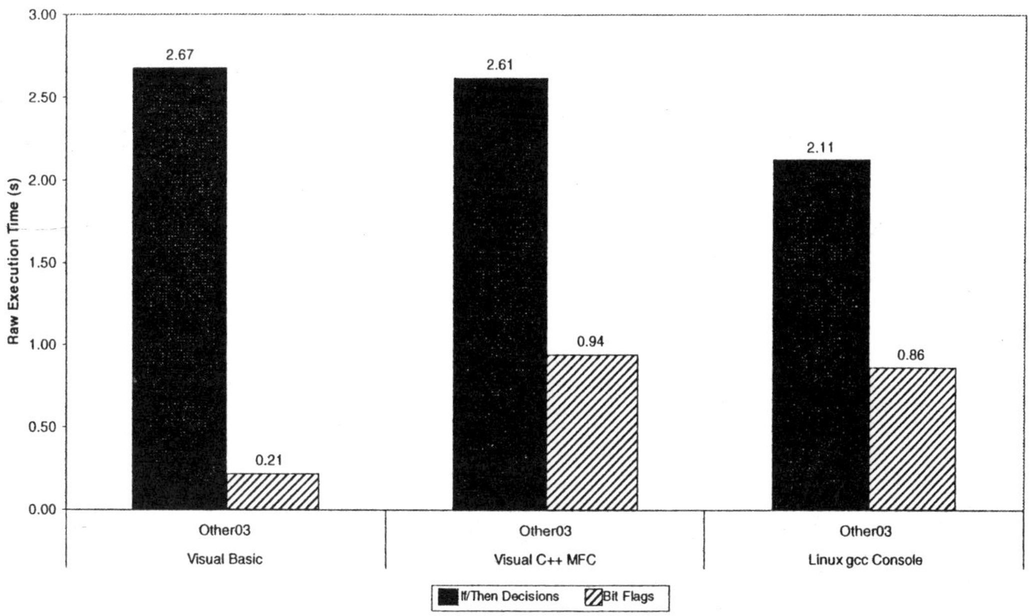

Figure 10.5 Demo *Other03* raw execution times for 50,000,000 iterations of an algorithm that must make a decision. An algorithm using a typical If ... Then ... Else structure is compared to one using bit flag encoding.

Chapter 11: Multi-tasking Basics

Though one may use highly optimized source code, compiled with optimized compilers, program execution ultimately reaches a point of maximum performance when running on a single computer running a single task. The answer to this performance wall may be parallel processing: breaking the problem into smaller pieces that can be solved concurrently. Before parallel processing is outlined in Chapter 12, however, a brief introduction of multi-tasking is presented in Chapter 11.

Multi-tasking can take many forms, and is perhaps a term often used without precise definition. The concepts of multi-tasking have significant consequences on the way a program interacts with the operating system, hardware and, in the case of some parallel architectures, other computers. In this chapter, some key terms related to multi-tasking and basic multi-tasking programming is presented.

11.1 MULTI-TASKING TERMS

The term multi-tasking itself warrants a basic definition and some discussion. Multi-tasking refers to the ability, or *apparent* ability, of a computer to run more than one program at one time. Strictly speaking, a single processor can only execute a single stream of instructions at any one time, but the *illusion* of multi-tasking can be made by relatively rapid context switching.

A context switch in a protected mode system occurs when the OS and hardware switch instruction streams from one program to another. This involves considerable overhead at the system level, as presented in Chapter 3. However, even with such overhead, a single processor machine gives the appearance of running multiple programs. Context switches can occur in a time-sliced system (each task gets a predetermined time) or interrupt system (a task makes requests for system resources). Generally, time slicing is used, but interrupts provide more real-time accuracy for time critical processes. The number of time slices given to a particular execution unit is determined by the *priority* of the unit. Priority is generally specified as a number, with higher priority units getting more time slices. For priorities, Windows also uses text descriptors such as "Normal" and "High." These descriptors correspond to numeric values on the priority scale.

In a thread based OS, such as Windows, the different "tasks" that are allocated time slices do not necessarily exist as separate application programs. Programs are typically begun as one or more process. A process is an execution unit that owns its own memory space in the system. Each process may spawn multiple threads, which are instruction streams that exist in the memory space of the creating process. That is, threads do not own their own memory, but share memory with any other threads in the process. This introduces a key concept in the consideration of multi-tasking: memory management in part determines the nature of a parallel system.

There are two broad types of memory management in parallel systems: shared memory or distributed memory. Multiple processes (whether on single or multiple processors) are examples of distributed memory execution units, whereas multiple threads within a process utilize shared memory. It is very important to define the memory model appropriately for a given problem, since many computational tasks are more suited to either distributed or shared memory.

To the scheduling portion of the OS, processes and threads are equivalent (and are collectively termed 'contexts'). That is, a process and a thread at the same priority each receive the same time slice. This is important to understand. Since threads get the same time slice and require fewer system resources to create and maintain, they are more efficient when properly used. For example, notice that in Figure 3.4, the number of threads created per second is far greater than the number of processes created per second.

Processes and threads are created by a creation process often called 'spawning;' a *parent* spawns a *child*, and the parent 'owns' the child. Parent processes can spawn child processes or threads, and threads can create processes or threads. The management of these units can get very tricky, to say the least! Generally, a process (or thread) can exist in one of four states: running, sleeping, dead or zombie. Running is an active state to the OS, and the code will get CPU resources. A sleeping execution unit is allocated memory, but has notified the OS that it does not require CPU time slices unless or until a 'waking signal' is received. A dead unit is one that has been 'killed,' and its resources are being released. Zombies cannot receive CPU resources, and are in a sense dead. Zombies continue to consume resources (usually memory), and may lead to memory leaks and other system problems.

Putting an execution unit to sleep is a valuable tool used to maintain system performance. A server may monitor a port for a request, but does not necessarily have to get a full time slice while idly listening. Giving a listening server a full time slice to 'do nothing' wastes CPU resources. Another pitfall of having the port monitored only during the time slice given to the process is lost requests. Therefore, it is not only a more efficient of use CPU resources but also more reliable if the OS kernel has facilities to 'continually' listen, then waking the server process only when a request is received. Such task manipulation requires considerable integration with the OS.As a practical example, consider the Apache http server running in Linux (there is a

Win32 Apache server, but it runs as a single process). When the server daemon is started, multiple processes are started (eight for a default configuration), and each child process may spawn multiple threads as needed (say 150). In this way, the system can respond to requests much faster than if a single process existed. If a request is already being handled by the server, the OS can wake the next process upon receiving a new request, rather than waiting for a current request to finish. This is illustrated in Figure 11.1. In Figure 11.1, a server request takes three active time slices to complete, the OS is running nine other active processes (not shown explicitly), and two requests arrive: one at t0 and the second at t20. The single process/thread model is shown in Figure 11.1(a). The server must complete the first request before it can respond to the second; the second request is not complete until time slice t50. In contrast, the second request is complete at the end of time slice t43 in the two process/thread model, as shown in Figure 11.1(b). Note that in Figure 11.1(b) there are 11 total threads getting time slices compared to ten in Figure 11.1(a). However, the performance gain remains. This savings can be significant in a heavily loaded system.

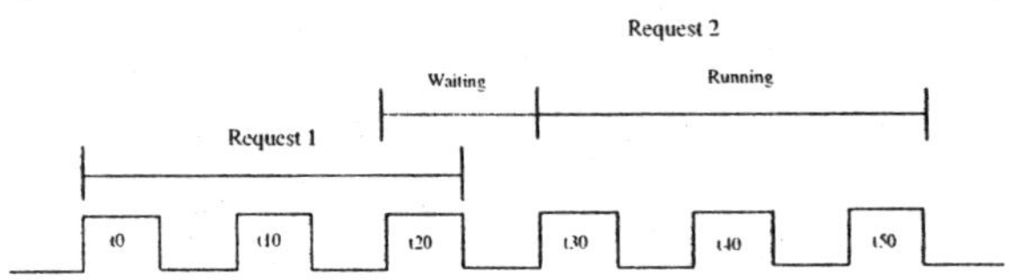

(a) Single Process/Thread

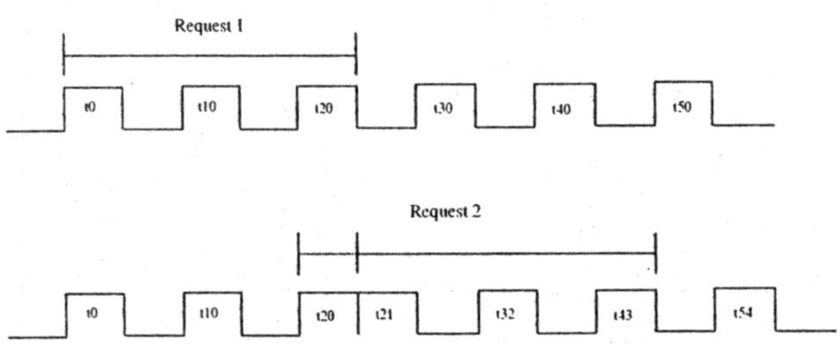

(a) Two Processes/Threads

Figure 11.1: Schematic diagram of processor resource allocation for a 'server' program running on a system with nine other processes (not shown, but they are given equivalent resource priority). The server receives two requests, at time slices t0 and t20. Figure 11.1(a) shows the server response in a single process (or thread) model. The two process (or thread) model is depicted in Figure 11.1(b).

11.2 MULTI-THREADED PROGRAMMING

Since execution threads are simpler units than processes, the practical application of threaded programming is presented first. This is somewhat arbitrary, and one may argue that process creation must occur first, since a process must exist in which to create the threads. However, the main process of a program is started by the OS when the program is executed. In other words, the programmer does not have to explicitly start the main process of a program.

A key point to recognize with multi-threaded applications is that even on single processor machines, they can give the *illusion* greater speed to the user though the actual execution time may be the same (or a little longer) when compared to a single thread version. Game programmers use this to considerable effect by drawing graphics in one thread and computing physics and artificial intelligence parameters in others. Assuming the drawing engine is 'fast enough,' the user sees a seamless series of frames drawn while the OS is continually switching from the drawing thread to the other compute threads. Figure 11.2 compares such a multi-threaded game to a single threaded version. Computational scientific applications that must supply continual graphical feedback to a user may benefit from a similar, multi-threaded approach. Specifically, the user sees graphical updates occurring *during* computational steps, rather than *after*. To the user, therefore, the code seems faster. A key point to note, however, is that the computational steps may take longer due to time sharing with the

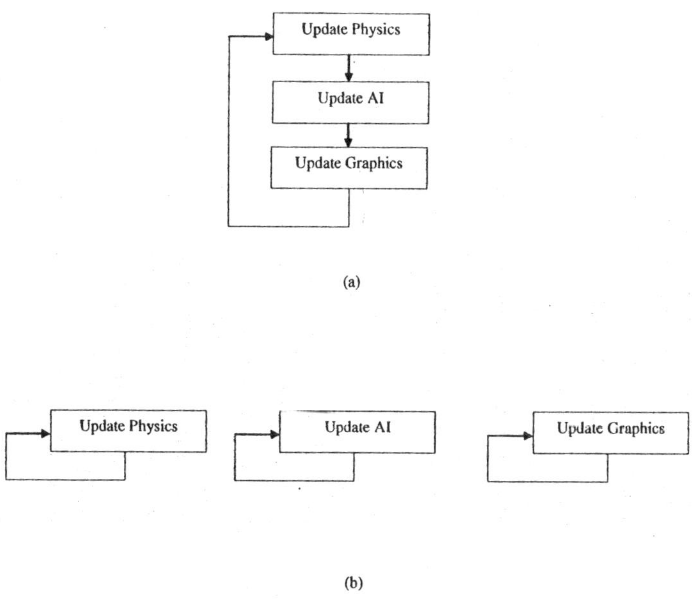

Figure 11.2: Simplified schematic diagram of a program with distinct, concurrent units (such as a game). In Figure 11.2(a), the program is a single thread. The multithread execution is shown in Figure 11.2(b).

drawing thread(s), but the user tends to not perceive this as much as static waiting for a new graphical frame.

To illustrate this, and a particular program that gives a *real* performance gain, Demo *Multi01* simulates a program with a graphical part and a compute part. This may be an actual working portion of code, or simply start-up initialization with a 'splash screen' as the graphics are initialized. The results of the Demo are presented after thread creation is discussed.

11.2.1 Simple Thread Creation and Termination

To demonstrate multi-threaded programming, the example uses a 'master' and 'worker' thread model. The master thread, the primary thread of the process, creates the worker thread. In a larger problem, the master may only perform administrative tasks, such as maintaining the worker threads. This is not a requirement in multi-threaded programming, but may be useful for thread management.

It should be noted that programming multi-threaded applications requires different headers and libraries than 'standard' programming. For example, the MS Visual C++ compiler has an option for using the multi-threaded libraries rather than single threaded ones, and with gcc, one uses –lpthread on the command line. In addition, it has been noted that MS VC++ compiler optimizations may not function properly in multi-threaded code. Therefore, it is very important to apply the programming techniques detailed in Section II rather than always relying on the compiler to do the optimization for you. Finally, full-featured multi-threaded programming is not supported in all languages. Visual Basic, for example, does support multi-threaded code, but uses the so-called 'apartment' memory model. This model does not allow different threads to share global (process-wide) data so that in effect, each thread is a contained unit. Multi-threaded applications can be written in BASIC, but this is by no means straightforward. Therefore, the demo's for this section are given only in C++.

Each process has at least one thread: the primary thread for that process. Creation of secondary (or child) threads, either from primary or child threads, is accomplished in Windows via a call to the Win API function `CreateThread`, which returns a handle to the created thread. The `CreateThread` function requires several parameters, which are outlined in Table 11.1 (Table 11.2 summarizes the use of the `pthread_create` function in Linux). Many of the parameters allow very fine control over how the thread is created and executes. These fine features are not generally needed for simple multi-threaded programming.

The thread itself is coded as a function *outside* the `Main` or `WinMain` functions. When the thread is created, a function name is passed to the `CreateThread` function (as shown in Table 11.1) that acts as a 'main' function for that thread. A given function can be called many times for as many threads are needed that use that function; each thread does not have to have a unique function. Further, a parameter

Table 11.1: Parameters of the Windows API CreateThread function.

Data Type	Parameter	Brief Description
LPSECURITY_ATTIBUTES	lpThreadAttributes	Pointer to a SECURITY_ATTRIBUTES structure; often, NULL is passed to accept the default security attributes
DWORD	dwStackSize	initial thread stack size; often, 0 is passed to accept the default stack size
LPTHREAD_START_ROUTINE	lpStartAddress	pointer to the thread function; generally, this is given as the function name itself
LPVOID	lpParameter	An argument for the new thread; this allows passing a single parameter to the thread itself.
DWORD	dwCreationFlags	creation flags; often, zero is passed.
LPWORD	lpThreadID	pointer to receive thread ID

Table 11.2: Parameters of the Linux pthread_create function.

Data Type	Parameter	Brief Description
pthread_t	thread	Pointer to thread id returned by pthread_create if successful
pthread_attr_t	attr	Pointer to attribute data; often, NULL will be passed to accept default attributes
void	start_routine (void *)	pointer to the thread starting function
void	arg	Pointer to arguments needed by thread; this can be single value, structure, etc.

can be passed to the thread, perhaps to control 'state' or some other behavior for the thread. That is, the thread can get one piece of data from a passed parameter; all other data the thread uses is global data. Finally, each thread has its own stack space (allocated within the memory of the creating process), so a thread can use its own set of local variables.

Though thread creation is relatively easy, properly terminating a thread is slightly more complex. First, the thread must actually be killed. There are three ways a thread can be killed. It can kill itself, be told to terminate (via a call to the `TerminateThread` Win API call) or be terminated when the process' primary thread terminates. To kill itself, the thread function (the one specified in the `CreateThread` call) simply completes and returns. To tell the thread to terminate can be done by a direct call to the API function, but this is not preferred, since the thread dies immediately without performing any exit code; memory leaks may result. A better way to tell a thread to die is tell it to kill itself. This way, the thread will perform the proper clean-up code upon exiting. Killing child threads by simply terminating the primary thread is also not preferred for the same reason.

Once a thread is dead, the handle to it must be released back to the Operating System. This is accomplished via a call to the `CloseHandle` function. In essence, killing a thread is a two-part process. The first step is the actual death of the thread, and the second, performed by a thread's parent, is informing the OS that the thread is dead. This allows the OS to reclaim all the resources that were used by the thread in a controlled way.

A simple demo application is presented in *Multi01* that compares a single threaded version to a multithreaded version. In this simplified demo, a 'worker' unit is simulated by a simple for loop, and a graphics unit is simulated by updates to a Progress Bar control on the user interface. Further, to simulate some load on the graphics unit, the function 'sleeps' for 25 milliseconds between updates. In the single thread version, the worker and graphics functions execute sequentially (in the main process thread). In the two-thread version, the worker unit is executed in a separate thread, and immediately after creation of that thread, the main process thread executes the graphics code. Raw total execution data for *Multi01* are presented in Figure 11.3.

There are two points about *Multi01* the merit specific discussion. First, in the single thread version, there is a noticeable delay before the graphics code is executed. This gives the user the 'appearance' of sluggishness in the program. Even if no total performance gain is realized by multi-threading these two units, by having the graphics updates begin immediately, without *waiting* for *completion* of the worker unit (that is, the graphics only has to wait for the next time slice), the user sees something happening and does not necessarily perceive the same degree of sluggishness. In

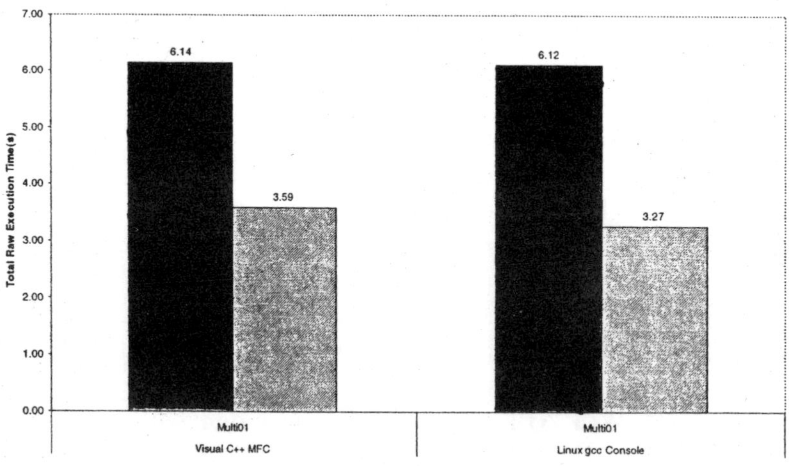

Figure 11.3: Demo *Multi01* raw execution times on the test computer for a single thread and two thread version of the same program. The total execution time (completion of both threads) is shown. Though this figure does show an overall performance increase for the two-thread version, this figure does not capture the perception of speed that multithreading can give, even in the absence of such a 'real' performance gain. Due to limitations in the multithreading implementation in Visual Basic, this demo was only done in C.

addition, for this particular demo, there is in fact a real, overall performance gain due to the time the graphics routine takes to execute. The worker unit can execute concurrently with the graphics updates (which ultimately depend on video hardware). As shown in Figure 11.3, the two-thread version executes about 1.7 times faster than the single thread, sequential execution of the two functions.

11.2.2 Communicating between threads

Except for very simple threaded applications, it will generally be necessary to pass information between threads. At the least, this includes communication between the primary process thread and any child threads that are created. Since threads share a common memory space, information may be shared in some circumstances by the use of global data (this is not allowed in the Apartment Model used in Visual Basic ActiveX COM objects), but care must be taken. There can be critical race situations between when a crucial piece of global data is read and modified by different threads. The OS does supply several data and code constructs such as semaphores, mutexes and critical sections that allow safer communication between threads. A key point here is that the OS supplies significant services in regard to thread management, and a multi-threaded program should generally take advantage of these system level 'tools' as needed.

Another key communication concern is ending the primary thread. Before a parent thread can safely either release child thread handles to the OS or terminate, it must 'know' that all its children have been properly terminated. Again, this is important to insure that all the required termination code in the children execute before the child threads are simply axed. The Windows API provides several functions to allow the primary thread to test the state of the children it has created. These functions test the *signal* of one or more threads. When a thread is running, it is *unsignaled* to the OS; however, when a thread terminates, it becomes *signaled* to the OS. The functions `WaitForSingleObject` and `WaitForMultipleObjects` are used to test for thread signals (with a thread handle and wait time passed as parameters). Simply put, a parent thread should test for signals from all its children before it calls either `CloseHandle` or exits. For gcc programs, one uses `pthread_join()` to wait for thread termination and release thread resources back to the OS.

11.2.3 When to use multiple threads

Since threads share memory space, they are useful whenever the shared memory model is appropriate. This is especially true when multiple threads must read a single large data array or structure, since only a single copy of the structure would be required. To use multiple processes in such a situation would require multiple copies of the data structure to be created; this either wastes physical system memory or increases communication overhead. Multiple threads can also be used to modify a very large data structure, especially when each thread can be allocated a particular portion of the data.

As shown in Figure 11.3, multiple threads can result in a performance increase even on single processor machines when certain threads have considerable latency. Since in a single processor machine, multiple threads do not execute simultaneously, distributing the processing of a large data structure across equivalent, multiple threads may seem like a waste of programming effort. However, Symmetric Multi-Processor (SMP) machines utilize shared memory, and it is with these computers that multi-threaded applications get a significant performance boost; the OS can allocate different threads to different processors. With each processor running a thread, crunching a particular part of the shared (global) data, concurrent processing occurs. This is an example of Multiple Instruction, Multiple Data (MIMD) parallel processing as discussed in Chapter 12. Therefore, writing a program using multiple threads introduces a favorable scalability if the code is migrated to a multiprocessor machine.

11.3 MULTI-PROCESS PROGRAMMING

An alternative to multi-threaded programming utilizes multiple processes. Each process has a unique memory space. Processes are 'higher order' entities to the OS and therefore require more OS overhead than threads. However, distributed systems cannot run multiple threads (except perhaps using highly specialized multi-processor OS kernels or distributed COM objects), with each thread sharing address space within one 'process.' Therefore, for programming systems with distributed memory spaces, multi-process is needed.

11.3.1 Simple Process Creation and Termination

Creating and terminating processes requires a bit more work for the OS than these operations require for threads. In Windows, the creation function is `CreateProcess`, and this function returns a Boolean value. As shown in Table 11.3, the parameters for the process creation function in the Windows API are more complex than those for thread creation (process creation in Linux is much simpler; one simply uses the `fork()` function to spawn a child process, then the child executes the desired program with the `execl()` function). A new process is created by specifying both the executable image and the command line needed for that program.

Killing a process is also more complex than killing a thread. There are several ways to kill a process programmatically, but Microsoft recommends the use of the `ExitProcess` function. This function performs some system clean-up in the form of notifying all process dependent dll's that the process is about to die; other means

Table 11.3: Parameters of the Windows API CreateProcess function

Data Type	Parameter	Brief Description
LPCTSTR	lpApplicationName	pointer to the name of the executable
LPTSTR	lpCommandLine	pointer to the command line string
LPSECURITY_ATTRIBUTES	lpProcessAttributes	pointer to structure holding process security attributes
LPSECURITY_ATTRIBUTES	lpThreadAttributes	pointer to structure holding thread security attributes
BOOL	bInheritHandles	handle inheritance flag
DWORD	dwCreationFlags	creation flags
LPVOID	lpEnvironment	pointer to new environment block
LPCTSTR	lpCurrentDirectory	pointer to new directory for the new process
LPSTARTUPINFO	lpStartupInfo	pointer to a startup structure that specifies how the process appears
LPPROCESS_INFORMATION	lpProcessInformation	pointer to structure that contains information about the process

of ending processes do not provide such clean-up, and may result in zombie dll's and leaks. When `ExitProcess` is called by a thread, all threads owned by the parent process are killed as well.

11.3.2 Communicating between Processes

Unlike multi-threaded applications, multiple processes do not contain a common memory space to store global data. For this reason, passing data between processes must be done via some explicit communication path between the processes. One method, though slow, is to use a disk file as the globally accessible data area. However, with this method, care must be taken in how the file is locked and unlocked for access, and that one process is not reading a partially modified file. In addition, any performance enhancement provided by disk caching is probably lost, since the file is constantly updated by the various processes. This method is convenient if only very limited data needs to be communicated (such as process ID's, run state, etc).

An alternative approach utilizes the network infrastructure provided by the OS. For example, one process can create a socket on which to 'listen' for the other process. Once the data path is established, each process can send data. Though the network infrastructure is used, this works even on a single machine. This is the paradigm used in many parallel programming systems, such as the Parallel Virtual Machine functions and the Message Passing Interface implementations.

11.3.3 When to Use Multiple Processes

Multiple processes are better suited to distributed memory applications than shared memory ones. Though multiple processes can be used for shared memory applications, communicating data between processes can introduce a large communication overhead. On the other hand, code that is to execute on different physical machines (that by definition have different memory spaces) has to be multiple process rather than multi-threaded. For example, a Beowulf parallel cluster requires each node to run a separate process, each with its own memory space.

Chapter 12: Parallel Computation Basics

When a particular program has been extensively optimized and single computer performance remains inadequate, parallel execution may be in order. Parallel computation is a rather significant leap from most of the coding conventions presented in Part II, and only a brief introduction is presented in this Chapter. The reader interested in (or requiring) parallel performance is encouraged to read the listed and similar references for specific information. Finally, in keeping the style of earlier chapters, only desktop computers are considered here for simplicity.

12.1 PARALLEL ARCHITECTURE BASICS

There are various techniques useful for parallel computation. Each has merits for particular tasks. Several approaches are briefly outlined in this section, and more detail is presented in Sections 12.2 and 12.3. There may be some overlap in these techniques, and others may exist.

12.1.1 Multiple Execution Units in the CPU

As discussed in Sections 2.5 and 2.7, many modern PC CPU's are designed with multiple execution units. These are implemented at the hardware level, with the CPU circuitry scheduling the decoded microinstructions for optimal performance. Performance can be further enhanced by the *order* in which CISC level instructions (as compiled from a high-level language) are presented to the CPU for decoding. The reader can imagine cases in which the order of source code lines do not matter logically for proper program execution; a point often overlooked, however, is that these 'equivalent' codes may not execute in the same number of CPU cycles due to scheduling within the CPU.

As an example, consider the two 3DNow! SIMD execution units used in the AMD Athlon processor. One execution unit is used in general for data moves and 'simple' operations such as addition. The other is utilized for 'complicated' operations such as multiplication and square roots. If from the same 'set,' sequential instructions must wait for the execution unit to become available (while the other sits idle). On the other hand, if the instructions are paired so that both execution units are utilized, two factors enhance performance. First, the CPU resources are better utilized. Second, instructions have less waiting time. A specific example of this optimization is presented in Section 12.2 after SIMD code is introduced.

12.1.2 Symmetric Multi-Processors (SMP)

Many high quality mainboards are available that allow installation of multiple processors. To effectively use the hardware in such systems, the OS must generally support multiple CPU's at the system level. This arrangement is particularly useful in multi-threaded or multi-process applications. SMP allows multiple instruction streams, and are most suited to shared memory systems (ie, all processors share the same physical memory space). SMP is a specific example of Multi Instruction Multiple Data parallelism discussed below.

12.1.3 Single Instruction, Multiple Data (SIMD)

The general x86 and x87 instructions, such as `ADD AX, 21`, can act on only one piece of data (21 in this example). Some computational problems are well suited to SIMD, which allows a single op-code instruction to act on multiple pieces of data. For example, signal or image processing applications are quite amenable to SIMD since the same operation often is performed on the entire data set. Imagine a lightening routine in an image processing application; a constant value is added to every element of what can be a very large array. Clearly, if the system can perform this operation with a single (or reduced number) instruction, performance improves, perhaps even if the latency for such an instruction is high. The specific SIMD implementation discussed in Section 12.2 is called SIMD Within A Register (SWAR).

12.1.4 Multiple Instruction, Multiple Data (MIMD)

Perhaps the most generally applicable parallel system is one composed of many execution units, each with their own memory space. Each unit in the system may actually be running completely different executables (or different functions within one large file). Applications for this architecture are numerous, and more details are presented in Section 12.3.

12.1.5 Algorithm Considerations

Deciding to solve a problem using parallel execution introduces several layers of complexity to the programming model. First, it must be determined if the problem is even suitable for concurrent processing. Moving a problem from serial to parallel execution involves several steps that are labeled and defined differently by different authors. The process presented here serves as an introduction; also noteworthy is that with parallel programming, there is generally more than one suitable approach.

If parallel execution can be applied to a problem, it will arise from one of two broad classes: functional or data concurrency. Functional concurrency arises in problems that numerous functions can be executed in parallel, with or without data dependencies. As a simple example, consider a game program. The program must compute various game object positions (which may be given by physics or user input, or both), collision detection, whatever artificial intelligence applies, etc. Many of these 'tasks' can be done in parallel if the resources exist. Multi-configuration *ab initio* quantum chemistry computations are suitable to such parallelism since each configuration can be computed on a separate node.

Data concurrency exists when the data set can be broken into smaller pieces. Many scientific applications can benefit from this type of parallelism, so long as data dependencies are properly handled. Some systems, however, may contain sufficiently strong dependencies that performance suffers. The ideal case, of course, is one for which the data subsets are completely independent.

Once the type of concurrency is determined, the actual process of implementing a parallel approach begins. The conceptual process can be described in three steps (adapted from Foster, *Designing and Building Parallel Programs,* which actually presents four steps in greater detail). This process does not include the details of actually implementing parallel code.

The first step is partitioning, which is essentially done in determining if the problem contains functional or data concurrency. While concurrency is the actual 'breakdown' of potential concurrent units, partitioning involves the programmatic act of dividing the problem: deciding *how* the functions or data will be subdivided into parallel units. Once this is done, the communications necessary must be delineated so that parallel processes remain synchronized and the full problem can be 'reassembled' at the end of the calculation. Finally, the partitioned problem with appropriate communications must be mapped to an actual computational architecture.

Once the problem has been mapped, the actual coding can begin. Of course, there are many tools and libraries available to streamline the implementation of parallel computation. In Section 12.4, several simple 'raw' approaches will be discussed; several tools that are available to simplify parallel program development are presented in Section 12.5.

12.1.6 Performance Considerations in Parallel Systems

The advantages of parallel computing do not come 'free.' In fact, this is yet another example of 'you can't get something for nothing, and you cannot break even.' A program that has the problem divided into n pieces for n nodes will likely not run in 1/n time of the single node computation, but will certainly execute faster than the single node time. Many parallel architectures exhibit a performance plateau, or even performance decreases, as the number of nodes increases. This introduces the idea of 'granularity,' the term for intrinsic concurrence of a problem. A formal definition of granularity, the compute time between communications, can be written as

(Total Compute Time) / (Number of Communications).

Granularity is generally described by three broad classes. First, there are 'embarrassingly parallel' problems which have so few data or functional dependencies that no communications are needed except for task creation and return of results. The granularity for these problems is essentially 'infinitely small.' Fine granularity problems have few communications so most effort goes to computation. These problems, typically algorithms with few data dependencies, scale 'favorably' with increasing the number of compute nodes. In contrast, with course granularity problems, data dependencies and the resulting communication overhead create a marked trade-off to increasing the number of compute nodes; due to the increased communications as number of nodes increases, course granularity problems reach of point of diminishing, or negative, return. This was clearly shown by Jayasimha, Hayder and Pillay in an evaluation of several parallel architectures for the same computational engineering problem (Navier–Stokes simulation of an axially symmetric jet). The granularity may be influenced by the problem itself or by the mapping to an actual computational platform.

Another consideration in performance is load balancing. In many systems, the compute nodes are homogeneous, so that as work units are parceled, it can be expected that each node perform the roughly same amount of work. On the other hand, heterogeneous systems may have some systems 'working hard' while other sit idle. Load balancing is also seen in parallel servers for which client requests are spread across the system. Load balancing increases the overhead requirements of the total system, but may improve overall performance.

12.2 SIMD

The x86 architecture has had an implementation of SIMD since the MMX instructions were added to the P5 core. These instructions allowed a single instruction to operate on multiple integer data. This resulted in significant performance improvements for the growing multimedia Internet applications that were becoming popular with the developing Pentium processor.

The implementation of MMX SIMD on the x86 chips involves a 'mode switch' that re-maps the FPU numeric registers to 64 bit MMX registers. Such a re-mapping precludes the simultaneous use of the registers as MMX and x87 numeric; the mode switch instruction clears the registers and separates blocks of MMX instructions from x87 instructions. Note, however, that x86 instructions can be used with either set. Depending on the processor, the mode switch may have considerable latency.

Therefore, the code should not repeatedly switch between modes.

12.2.1 Integer Array Addition with MMX

As stated, the MMX instructions operate on integer data. A single MMX register, being 64 bits wide, can hold up to 8 8-bit integers, 4 16-bit integers or 2 32-bit integers. Suppose **a** and **b** are eight element byte arrays (equivalent to 8-dimensional vectors). Using the x86 scalar instructions in an unrolled algorithm, addition of **c** = **a** + **b** would require element-by-element addition like the one shown in **Listing 12.1.**

Listing 12.1 ASSEMBLY Language listing for x86 addition of two 8 element byte arrays

```
MOV        EAX, a[1]
MOV        EDX, b[1]
ADD        EAX, EDX
MOV        c[1], EAX
MOV        EAX, a[2]
MOV        EDX, b[2]
ADD        EAX, EDX
MOV        c[2], EAX
MOV        EAX, a[3]
MOV        EDX, b[3]
ADD        EAX, EDX
MOV        c[3], EAX
MOV        EAX, a[4]
MOV        EDX, a[4]
ADD        EAX, EDX
MOV        c[4], EAX
MOV        EAX, a[5]
MOV        EDX, b[5]
ADD        EAX, EDX
MOV        c[5], EAX
MOV        EAX, a[6]
MOV        EDX, b[6]
ADD        EAX, EDX
MOV        c[6], EAX
MOV        EAX, a[7]
MOV        EDX, b[7]
ADD        EAX, EDX
MOV        c[7], EAX
```

```
MOV         EAX,  a[8]
MOV         EDX,  b[8]
ADD         EAX,  EDX
MOV         c[8],  EAX
```

The code in **Listing 12.1** requires 32 cycles on the Pentium. In contrast, the MMX instructions can be used as shown in **Listing 12.2**.

Listing 12.2 8 element array addition using MMX SIMD instructions

```
EMMS                            ; clears registers for MMX use
MOVQ        MM0,  a[0]          ; MM0 holds eight bytes of a
MOVQ        MM1,  b[0]          ; MM1 holds eight bytes of b
PADD        MM0,  MM1           ; MM0 holds eight bytes a + b
MOVQ        c[0],  MM0          ; eight bytes of result put into
                                ; memory
EMMS                            ; clears registers, okay for x87
                                ; use
```

In **Listing 12.2**, the arrays are actually added with a *single* instruction: PADD (for "Packed ADD"). **Listing 12.2** can execute in 4 cycles, not counting the two EMMS instructions! At one cycle per packed add instruction, this is a definite improvement.

12.2.2 Floating Point SIMD and Testing for SIMD Capability

While the MMX instructions themselves act only on integer data, SIMD floating-point support was added to Intel processors with the SSE instruction set extension (a subset of SSE implemented by AMD was termed 3DNow!). SSE introduced a new set of 128 bit registers, while the 3DNow! instructions use the existing 64-bit MMX registers. Each of these SIMD extensions take single precision floating point operands, so a single MMX 64-bit register can hold two 32-bit operands. Therefore, two single precision floating-point operations can be accomplished with a single 3DNow! instruction. Before using the SIMD floating-point instructions (or MMX integer instructions, for that matter), the CPU capability should be checked (after testing for support of the `CPUID` instruction). A routine to test for 3dNow! support is given in the ***AMD 3DNow! Technology Manual*** and is shown in **Listing 12.3a**, and the analogous routine for testing for SSE support is shown in **Listing 12.3b** (these two code snippets assume that the `CPUID` instruction is supported; there is no explicit test for `CPUID` support shown). The specific implementation of this code for inline ASSEMBLY can be seen in Demos *Par01*, *Par02* and *Par03*.

Listing 12.3a Assembly Language routine to test for 3DNow! support

```
xor   eax, eax          ; CPUID function: Vendor
                        ; ID
CPUid                   ; Invoke CPUid function
test  eax, eax
jz    $No3dNow
mov   eax, 80000000h    ; CPUID function: Largest
                        ; extended value
CPUid
cmp   eax, 80000001h    ; We can execute feature
                        ; #1 right?
jl    $No3dNow          ; If not, 3DNow! is not
                        ; supported
mov   eax, 80000001h    ; CPUID function:
                        ; Signature + features
CPUid
test  edx, 80000000h    ;test bit 31
jnz   YES_3Now!         ;3DNow! is supported
```

Listing 12.3b Assembly Language routine to test for SSE support

```
            push eax
            push edx
            xor  eax, eax          ; CPUid function VendorID
            CPUid
            test eax, eax
            jz   $NoSSE
            mov  eax, 00000001h    ; CPUid function Largest
                                   ; Ext. Val
            CPUid
            test eax, 02000000h
            jl   $YesSSE
            jmp  $NoSSE
$YesSSE:    mov  bSSEOK, 1         ; bSSEOK = true
            jmp  $EndCpuTest
$NoSSE:     mov  bSSEOK, 0         ; bSSEOK = false
$EndCpuTest:     pop  edx
            pop  eax
```

Any program that uses extended, CPU specific, instructions written for execution on an unknown computer should test for support of the extended functions. However, the program should be structured in such a way that the test need only be performed *once*, rather than each time SIMD instructions are needed. Once the CPU instruction set is determined at run time, a function pointer can be used to call the appropriate functionality.

The IBM processors also have SIMD capability. The MPC7400, for example, implements SIMD in a way that differs from the Intel and AMD methods in two key respects. First, the register depth for the 128 bit SIMD registers is considerably deeper; there are 32 128 bit SIMD registers on the 7400! In addition, the 7400 has Altivec, a dedicated vector processing unit. This is in direct contrast to the AMD 3dNow! implementation for which the vector math unit is shared with the regular scalar floating point unit.

12.2.3 Vector Dot Product using SIMD

As a computational example of 3DNow!, consider the three dimensional vector dot product **c** = **a** · **b**. Recall from Chapter 7 that vectors such as **a** and **b** are conveniently stored in arrays (which to be used with SSE/3DNow! instructions, must be single precision), and two algorithms for the dot product were presented in Section 7.1. In x87 code, the unrolled algorithm can be coded in inline ASSEMBLY Language as shown in **Listing 12.4** (which is the ASSEMBLY listing from compiled C).

Listing 12.4 **x87 based dot product of two single precision 3-d vectors**

```
FLD         a[0]        ; indexes here are actual
                        ; byte offsets, st(0) is
                        ; implicit
FMUL        b[0]
FLD         a[4]
FMUL        b[4]
FADDPST(1), ST(0)
FLD         a[8]
FMUL        b[8]
FADDPST(1), ST(0)
FSTP        c
```

The code in **Listing 12.4** requires 26 cycles on an Athlon. In contrast, the SIMD approach is shown in **Listing 12.5**.

Listing 12.5a **3dNow! SIMD dot product of two 4-d vectors**

```
movq        mm0, a[0]       ; mm(0) = a(0) | a(1)
movq        mm3, b[0]       ; mm(1) = b(0) | b(1)
movq        mm2, a[8]       ; don't use movd!!  SLOW!
movq        mm1, b[8]
pfmulmm0,   mm3             ; mm(0) = a(0)*b(0) | a(1) * b(1)
pfmulmm1,   mm2
pfaccmm0,   mm0
pfaddmm0,   mm1
movd        c, mm0          ; c = mm(0)lo
```

Listing 12.5b **SSE SIMD dot product of two 4-d vectors**

```
movaps      xmm0, a[0]
movaps      xmm1, b[0]
mulpsxmm0,  xmm1
movaps      xmm1, xmm0
shufps      xmm1, xmm0, 0xee
addpsxmm1,  xmm0
movaps      xmm0, xmm1
shufps      xmm1, xmm1, 0xe5
addpsxmm1,  xmm0
movssxmm0,  c
```

Listing 12.5a also executes in 26 cycles on an Athlon, again ignoring the 'mode switch' instructions (which has a latency of 2 cycles on the Athlon). Though a lack of performance gain exists for the SIMD code shown in **Listing 12.5**, an improvement can be made by more effectively utilizing *both* 3dNow! instruction pipes. For example, the optimized dot product routine from the AMD SDK library is shown in **Listing 12.6**.

Listing 12.6 **3dNow! Dot product routine optimized to utilize instruction scheduling**

```
movq        mm0, a[0]       ; mm(0) = a(0) | a(1)
movq        mm3, b[0]       ; mm(1) = b(0) | b(1)
pfmulmm0,   mm3             ; mm(0) = a(0)*b(0) | a(1) * b(1)
movq        mm2, a[8]       ; don't use movd!!  SLOW!
movq        mm1, b[8]
pfaccmm0,   mm0
pfmulmm1,   mm2
pfaddmm0,   mm1
movd        c, mm0          ; c = mm(0)lo
```

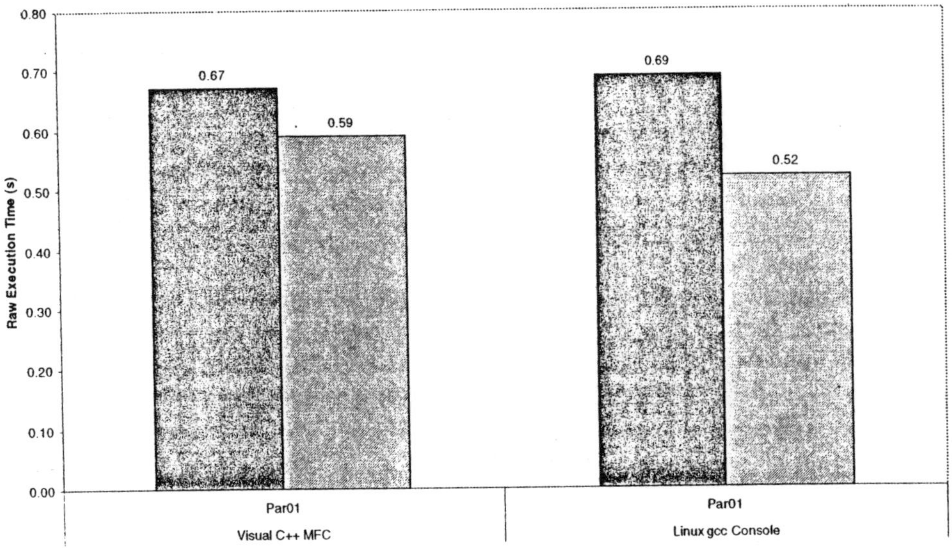

Figure 12.1: Demo *Par01* raw execution times for 100,000,000 iterations of the 3-d vector dot product; the test computer (1200 MHz AMD Athlon) used as the test platform for the earlier chapters was used. Compared are C compiled to x87 instructions and inline ASSEMBLY 3dNow! SIMD instructions. Note that the SIMD routine works on 4-d vectors (the fourth dimension is 'free' and faster).

Execution performance for **Listing 12.6**, compared to the simple, scalar unrolled version exemplified in **Listing 7.3,** is demonstrated by Demo *Par01*. Raw execution times for *Par01*, shown in Figure 12.1, indicate that the properly scheduled version runs about 12% faster. Actually, *Par01* uses '4-d' vectors. The fourth element is set to 0 to perform a 3-d dot product.

12.2.4 4x4 Matrix Multiplication using SIMD

Many algorithms can be coded to take advantage of the SIMD instructions. Another specific example from Part II is matrix multiplication: **C** = **A** x **B**. However, the nature of the matrix multiplication introduces a slight complication in the handling of this operation. Recall from Chapter 5 that multidimensional arrays are typically stored in 'row-major' order, with the elements mapped to memory as shown in Figure 5.6. Since a MOVQ instruction moves two elements from consecutive memory, it is impossible to move, say, b[0][0] and b[1][0] with a single MOVQ with a row-major array. Of course, one could use the MOVD instruction, followed by PUNPCKHDQ; however, this method is slower than MOVQ since an extraneous instruction is needed (the PUNPCKHDQ) and on older processors, the MOVD has greater latency than MOVQ. Further, this would require *two* MOVD for the two pieces of data (which is more of a scalar move). A better approach, harking back to the point that for true high performance code careful data planning is needed, is to store the second matrix (**B** in this case) in 'column-major' order, which requires the storage of $\mathbf{B}^T$. Using the second matrix stored in this way allows the use of the relatively efficient MOVQ instructions to get the second matrix data into MMX registers. Assuming **B** is stored 'column-major,' with the memory mapping analogous to that shown in Figure 12.2, the code shown in **Listing 12.7a** can be used to compute the 4x4 matrix multiplication.

An efficient SSE 4x4 matrix multiplication algorithm treats the **A** matrix as separate row vectors, and the entire **B** matrix is loaded into SSE registers. This implementation of the matrix multiplication does not require storing of **B** in column major order. The code for SSE 4x4 matrix multiplication is shown in **Listing 12.7b**.

Figure 12.2: Memory map of a simple 2x2 array stored in Column Major order. The array, a, is single precision, so each element is four bytes.

Memory Address	Data Value
1000 – 1003	a(0,0)
1004 – 1007	a(1,0)
1008 – 1011	a(0,1)
1012 – 1015	a(1,1)

Listing 12.7a 3DNow! 4x4 matrix multiplication

```
;compute c_00-c_03
MOVQ        MM0,  a[0]        ;mm0=  a_00|a_01
MOVQ        MM1,  a[8]        ;mm1=  a_02|a_03
MOVQ        MM2,  b[0]        ;mm2=  b_00|b_10
MOVQ        MM3,  b[8]        ;mm3=  b_20|b_30
PFMULMM2,   MM0               ;mm2=  a_00*b_00|a_01
                              ;  *b_10
MOVQ        MM4,  b[16]       ;mm4=  b_01|b_11
MOVQ        MM5,  b[24]       ;mm5=  b_21|b_31
PFMULMM3,   MM1               ;mm3=  a_02*b_20|a_03
                              ;  *b_30
MOVQ        MM6,  b[32]       ;mm6=  b_02|b_12
MOVQ        MM7,  b[40]       ;mm7=  b_22|b_32
PFMULMM4,   MM0               ;mm4=  a_00*b_01|a_01
                              ;  *b_11
```

```
PFADDMM2,  MM3          ;these next two steps
                              ;form
PFACCMM2,  MM2          ;c_00, the sum of
                              ;products
MOVQ       c[0],  MM2         ;store c_00
PFMULMM5,  MM1          ;mm5= a_02*b_21|a_03
                              ;*b_31
MOVQ       MM3, b[56];mm3= b_23 | b_33
MOVQ       MM2, b[48];mm2= b_03 | b_13
PFMULMM6,  MM0          ;mm6= a_00*b_02|a_01
                              ; *b_12
PFADDMM4,  MM5          ;add products to form
PFACCMM4,  MM4          ;c_01
MOVQ       c[4],  MM4         ;store c_01
PFMULMM7,  MM1          ;mm7= a_02*b_22|a_03
                              ; *b_32
PFADDMM6,  MM7          ;add products to form
PFACCMM6,  MM6          ;c_02
MOVQ       c[8],  MM6         ;store c_02
PFMULMM2,  MM0          ;mm2= a_00*b_03|a_01
                              ; *b_13
MOVQ       MM0, a[16];mm0= a_10 | a_11
PFMULMM3,  MM1          ;mm3= a_02*b_23|a_03
                              ; *b_33
MOVQ       MM1, a[24];mm0= a_12 | a_13
PFADDMM2,  MM3          ;add products for
PFACCMM2,  MM2          ;form c_03
MOVQ       c[12], MM2;store c_03

;compute c_10-c_13
;comments as for c_00-c03, except contents of
;mm0&mm1
MOVQ       MM2, b[0]
MOVQ       MM3, b[8]
PFMULMM2,  MM0
MOVQ       MM4, b[16]
MOVQ       MM5, b[24]
PFMULMM3,  MM1
MOVQ       MM6, b[32]
MOVQ       MM7, b[40]
PFMULMM4,  MM0
PFADDMM2,  MM3
PFACCMM2,  MM2
MOVQ       c[16], MM2
PFMULMM5,  MM1
MOVQ       MM3, b[56]
```

```
MOVQ        MM2,  b[48]
PFMULMM6,   MM0
PFADDMM4,   MM5
PFACCMM4,   MM4
MOVQ        c[20],  MM4
PFMULMM7,   MM1
PFADDMM6,   MM7
PFACCMM6,   MM6
MOVQ        c[24],  MM6
PFMULMM2,   MM0
MOVQ        MM0,  a[32]
PFMULMM3,   MM1
MOVQ        MM1,  a[40]
PFADDMM2,   MM3
PFACCMM2,   MM2
MOVQ        c[28],  MM2

;compute c_20-c_23
;comments as for c_00-c03, except contents of
;mm0&mm1
MOVQ        MM2,  b[0]
MOVQ        MM3,  b[8]
PFMULMM2,   MM0
MOVQ        MM4,  b[16]
MOVQ        MM5,  b[24]
PFMULMM3,   MM1
MOVQ        MM6,  b[32]
MOVQ        MM7,  b[40]
PFMULMM4,   MM0
PFADDMM2,   MM3
PFACCMM2,   MM2
MOVQ        c[32],  MM2
PFMULMM5,   MM1
MOVQ        MM3,  b[56]
MOVQ        MM2,  b[48]
PFMULMM6,   MM0
PFADDMM4,   MM5
PFACCMM4,   MM4
MOVQ        c[36],  MM4
PFMULMM7,   MM1
PFADDMM6,   MM7
PFACCMM6,   MM6
MOVQ        c[40],  MM6
PFMULMM2,   MM0
MOVQ        MM0,  a[48]
PFMULMM3,   MM1
```

```
MOVQ        MM1,  a[56]
PFADDMM2,   MM3
PFACCMM2,   MM2
MOVQ        c[44],  MM2

;compute c_30-c_33
;comments as for c_00-c03, except contents of
;mm0&mm1
MOVQ        MM2,  b[0]
MOVQ        MM3,  b[8]
PFMULMM2,   MM0
MOVQ        MM4,  b[16]
MOVQ        MM5,  b[24]
PFMULMM3,   MM1
MOVQ        MM6,  b[32]
MOVQ        MM7,  b[40]
PFMULMM4,   MM0
PFADDMM2,   MM3
PFACCMM2,   MM2
MOVQ        c[48],  MM2
PFMULMM5,   MM1
MOVQ        MM3,  b[56]
MOVQ        MM2,  b[48]
PFMULMM6,   MM0
PFADDMM4,   MM5
PFACCMM4,   MM4
MOVQ        c[52],  MM4
PFMULMM7,   MM1
PFADDMM6,   MM7
PFACCMM6,   MM6
MOVQ        c[56],  MM6
PFMULMM2,   MM0
PFMULMM3,   MM1
PFADDMM2,   MM3
PFACCMM2,   MM2
MOVD        c[60],  MM2
```

Listing 12.7b SSE 4x4 matrix multiplication

```
; load B matrix into SIMD registers
movaps       xmm4,  b[0]
movaps       xmm5,  b[16]
movaps       xmm6,  b[32]
movaps       xmm7,  b[48]

;iterate through A matrix as row vectors
```

```
;using unrolled loop

; load first row vector into reg 0-3
; and broadcast
movaps    xmm0, a[0]
movaps    xmm1, xmm0
movaps    xmm2, xmm0
movaps    xmm3, xmm0
shufps    xmm0, xmm0, 0x00
shufps    xmm1, xmm1, 0x55
shufps    xmm2, xmm2, 0xaa
shufps    xmm3, xmm3, 0xff

;multiply the elements
mulps     xmm0, xmm4
mulps     xmm1, xmm5
mulps     xmm2, xmm6
mulps     xmm3, xmm7

;add the products and store result
addps     xmm1, xmm0
addps     xmm2, xmm3
addps     xmm1, xmm2
movaps    c[0], xmm1

; load second row vector into reg 0-3
; and broadcast
movaps    xmm0, a[16]
movaps    xmm1, xmm0
movaps    xmm2, xmm0
movaps    xmm3, xmm0
shufps    xmm0, xmm0, 0x00
shufps    xmm1, xmm1, 0x55
shufps    xmm2, xmm2, 0xaa
shufps    xmm3, xmm3, 0xff

;multiply the elements
mulps     xmm0, xmm4
mulps     xmm1, xmm5
mulps     xmm2, xmm6
mulps     xmm3, xmm7

;add the products and store result
addps     xmm1, xmm0
addps     xmm2, xmm3
addps     xmm1, xmm2
```

```
movaps     c[16],  xmm1

; load third row vector into reg 0-3
; and broadcast
movaps     xmm0,  a[32]
movaps     xmm1,  xmm0
movaps     xmm2,  xmm0
movaps     xmm3,  xmm0
shufps     xmm0,  xmm0,  0x00
shufps     xmm1,  xmm1,  0x55
shufps     xmm2,  xmm2,  0xaa
shufps     xmm3,  xmm3,  0xff

; multiply the elements
mulps      xmm0,  xmm4
mulps      xmm1,  xmm5
mulps      xmm2,  xmm6
mulps      xmm3,  xmm7

;add the products and store result
addps      xmm1,  xmm0
addps      xmm2,  xmm3
addps      xmm1,  xmm2
movaps     c[32],  xmm1

; load fourth row vector into reg 0-3
; and broadcast
movaps     xmm0,  a[48]
movaps     xmm1,  xmm0
movaps     xmm2,  xmm0
movaps     xmm3,  xmm0
shufps     xmm0,  xmm0,  0x00
shufps     xmm1,  xmm1,  0x55
shufps     xmm2,  xmm2,  0xaa
shufps     xmm3,  xmm3,  0xff

;multiply the elements
mulps      xmm0,  xmm4
mulps      xmm1,  xmm5
mulps      xmm2,  xmm6
mulps      xmm3,  xmm7

;add the products and store result
addps      xmm1,  xmm0
addps      xmm2,  xmm3
addps      xmm1,  xmm2
movaps     c[48],  xmm1
```

The code in **Listing 12.7**, implemented in Demo *Par02*, may not be fully optimized with regard to instruction pairing and most efficient register use, but does execute approximately 1.6 times faster than the unrolled (and compiled to x87 op-codes) code as used in demo *Loop03*. Raw execution times for Demo *Par02* are shown in Figure 12.3.

One thing is clear from **Listing 12.7a**: even the modest sized 4x4 matrix multiplication suffers from MMX register deficiency. Additional registers would allow the storage of the entire matrix **B** data without the repeated moves. L1 caching is the next best option, and since the matrix **B** is used for every element of **C** that is computed, a processor with are larger L1 data cache will have an advantage for matrix multiplication.

12.2.5 Simpson's Rule Integration using SIMD

The partially unrolled Simpson's Rule Integration, also from Chapter 7, is another algorithm that can benefit from vector instructions. It should be noted, however, like the matrix multiplication example above, this comes at the expense of a counter-intuitive storage of the data. In Demo *Loop04*, the function was stored in a straightforward manner: g_0 mapped to array element f[0], g_1 mapped to array element f[1]. etc. To take advantage of the SIMD instructions, which work with *successive* values in memory, f[2n] and f[1] need to be switched. Recall from Chapter 7 the three main sums needed for the integration are:

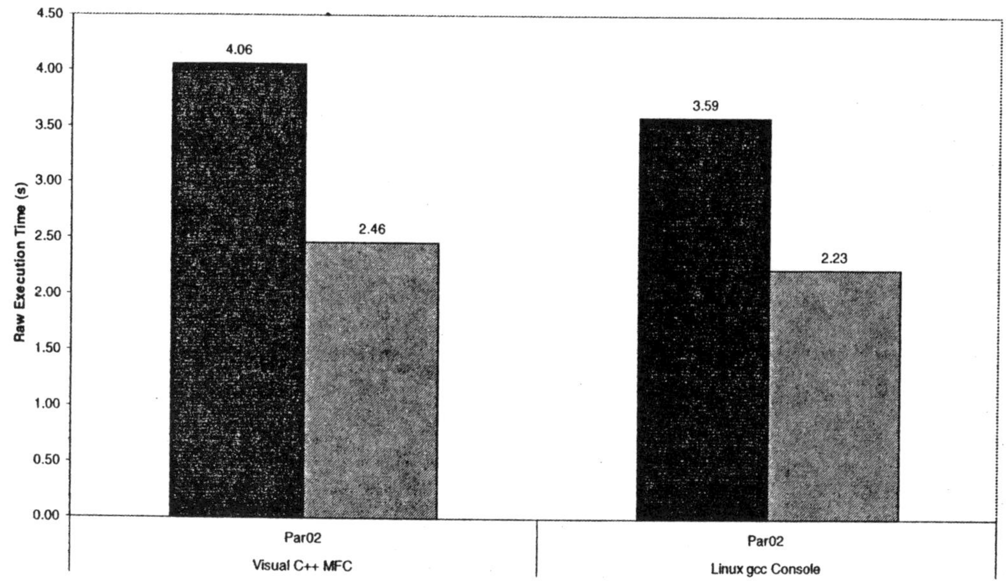

Figure 12.3: Demo *Par02* raw execution times on the test computer for 50,000,000 iterations of a 4 x 4 matrix multiplication. Compared are C compiled to x87 instructions and inline ASSEMBLY 3dNow! SIMD instructions.

Eq. (12.1) $s_0 = g_0 + g_{2n}$
Eq. (12.2) $s_1 = g_1 + g_3 + \dots g_{2n-1}$
Eq. (12.3) $s_2 = g_2 + g_4 + \dots g_{2n-2}$

It is impossible to store g_0 and g_{2n} in an SIMD register with a single instruction (again, the `movd`, `punpckldq`, `movd` combination can be used, but this is relatively slow). By switching the array elements for g_1 and g_{2n}, each of the three sums in Eq. (12.1) – (12.3) can be computed in SIMD registers. Also, notice that the 2 and 4 factors for s_1 and s_2 are stored as a 'vector' so they can be used in an SIMD instruction. The 'first run' sample code (not optimized) is shown in **Listing 12.8**.

Listing 12.8a Partially unrolled Simpson's Rule Integration using 3dNow! SIMD

```
//switch f[1] and f[n]
//this is needed to access first and last as
//a single eight byte 'vector'
tempval  = f[1];
f[1]  = f[9999];
f[9999]  = tempval;

//initialize the factors
factvec[0]  = 2;
factvec[1]  = 4;

//h/3, where h=(b-1)/n
//moved out of loop to avoid mmx vs fp
//register collision
h_3  = (float)0.999 /30000;

// start the assembly routine
_asm femms


      //initialize mm0 with f[2]|f[3] and mm3
      //with h_3|h_3
      //eax is pushed in case it is being
      //used for something important
      __asm {
                  movq        mm0,  f[8]
                  movd        mm3,  h_3
                  punpckldq   mm3,  mm3
                  push        eax


      }
```

```
//do the summation unrolled 8 terms
//each for s1 and s2
for(SimpCounter = 4; SimpCounter <= 9974;
      SimpCounter = SimpCounter + 16){

    __asm{

        mov               eax, SimpCounter
        pfadd      mm0,  f[eax * 4]
        pfadd      mm0,  f[8 + eax * 4]
        pfadd      mm0,  f[16 + eax * 4]
        pfadd      mm0,  f[24 + eax * 4]
        pfadd      mm0,  f[32 + eax * 4]
        pfadd      mm0,  f[40 + eax * 4]
        pfadd      mm0,  f[48 + eax * 4]
        pfadd      mm0,  f[56 + eax * 4]

    }


//clean up extra terms
_asm{

    pfadd      mm0,  f[39952]
    pop        eax
    pfadd      mm0,  f[39960]
    pfadd      mm0,  f[39968]
    pfadd      mm0,  f[39976]
    pfadd      mm0,  f[39984]
    pfadd      mm0,  f[39992]

}


_asm{

    movq       mm1,  factvec
    pfmulmm0,  mm1
    movq       mm2,  f[0]
    pfaddmm0,  mm2
    pfmulmm0,  mm3
    pfaccmm0,  mm0
    movd       Integral, mm0
}
```

```
_asm femms
```

Listing 12.8b Partially unrolled Simpson's Rule Integration using SSE SIMD

```
// for the SSE version, we need some additional
// vectors to avoid the scalar operations at the end
// initialize those first

//initialize factor vector
factvec[0] = 4;
factvec[1] = 2;
factvec[2] = 4;
factvec[3] = 2;

//initialize partialh_3
partialvec[0] = 1;
partialvec[1] = 1;
partialvec[2] = 4;
partialvec[3] = 2;

//initialize the zero bytes
zerobytes[0] = 0;
zerobytes[1] = 0;
zerobytes[2] = 0;
zerobytes[3] = 0;


//switch f[1] and f[n]
//this is so f[0] and f[n] are in sequential bytes
//to be accessed as a vector
tempval = f[1];
f[1] = f[9999];
f[9999] = tempval;

//h/3, where h=(b-a)/n
//moved out of loop to avoid mmx vs fp register collision
h_3 = (float)0.999 /30000;


// begin the assembly routine
_asm {
```

```
        ;preload f[4]..f[7] into xmm0
        movaps    xmm0, f[16]

        ;broadcast h_x into xmm3
        movss     xmm3, h_3
        shufps    xmm3, xmm3, 0x00

        ;load h_3 into high dword of xmm7
        ;1 into low dword of xmm7
        movaps      xmm7, partialvec

        ;save eax
        push eax

        ;zero accumulation registers
        movaps      xmm4, zerobytes[0]
        movaps      xmm5, zerobytes[0]
        movaps      xmm6, zerobytes[0]

}

//do the partially unrolled summation
for(SimpCounter = 8; SimpCounter <= 9976; SimpCounter =
SimpCounter + 16){

        _asm {


                ;accumulate sum for this iteration
                addps    xmm0, f[eax * 4]
                addps    xmm4, f[16 + eax * 4]
                addps    xmm5, f[32 + eax * 4]
                addps    xmm6, f[48 + eax * 4]
        }

}

_asm {

;clean up terms not in sum
addpsxmm0, f[0x2708 * 4]
addpsxmm5, f[16 + 0x2708 * 4]

;accumulate into mm0
;and restore eax
addps    xmm0, xmm4
```

```
addps     xmm6,  xmm5
pop         eax
addps     xmm0,  xmm6

;compute products and complete sums of S0, S1, S2

movaps    xmm2,  f[0]
movaps      xmm4,  factvec[0]
mulps     xmm0,  xmm4
mulps     xmm2,  xmm7
addps     xmm0,  xmm2
mulps     xmm0,  xmm3

;accumulate xmm0 to a scalar result
movaps    xmm1,  xmm0
shufps    xmm1,  xmm0,0xee
addps     xmm1,  xmm0
movaps    xmm0,  xmm1
shufps    xmm1,  xmm1,0xe5
addps     xmm1,  xmm0
movss     Integral,  xmm1

}
```

Raw execution times for Demo *Par03*, which compares the fastest (x87 only) routine from Demo *Loop04* to the SIMD version in **Listing 12.8**, are shown in Figure 12.4. The 3dNow! vector code ran about two times faster on Windows with Visual C++, but not on Linux with gcc (which may be due to different compiler optimizations).

12.2.6 A Comparison of 3dNow! to SSE SIMD Performance

The AMD Athlon processors (beginning with the Palomino core) have SSE as well as 3dNow! capability and present an interesting opportunity to compare the SSE instruction performance to the 3dNow! technology. This is a better test of these two instruction sets than is seen in many 'benchmark' tests that run on separate processors, which therefore by definition also have different main board chipsets and other system components. To examine how the specific AMD implementation of SSE affects the results, the tests were also done on a Pentium III computer. Each of these comparisons is for Linux gcc console applications compiled with the `-O1` compiler switch and `-march=<athlon or pentiumpro>`. Figure 12.5 summarizes the performance results for *Par01* – *Par03* running on three architectures.

Reduced empirical clock cycles per iteration for Demo *Par01* comparing x87, 3dNow! and SSE running an Athlon (Thunderbird) 1200, an Athlon XP 2100+ (1733 MHz, Palomino core) and a 650 MHz Pentium III are shown in Figure 12.5a. Note that not only is the SSE algorithm in this case slower than the 3dNow!, but SSE is also slower than the scalar x87 algorithm. The relative weakness of the SSE instruction set for performing dot products is the absence of a packed accumulate (similar to the PFACC in 3dNow!) instruction in SSE. In essence, the final addition of the component products must be done by scalar addition after shuffling the packed data (for a 3-d dot product, there are three multiplies and two adds, so the scalar addition is roughly half the total algorithm). Such inefficient use of the vector processor to perform a scalar addition causes a significant performance hit.

Demo *Par02* results for the Athlon Thunderbird, Athlon Palomino, and Pentium processors are shown in Figure 12.5b. Interestingly, in this case, the SSE again performs more poorly than the both 3dNow! and x87 code on the Athlon Palomino. However, the Pentium implementation of SSE is three times faster than the x87 scalar code. This shows that the SSE implementations are quite different for the Intel and AMD processors. This conclusion has significant consequences when writing code for the general processor and SSE is chosen; in the case of the AMD processors, SSE is not the better choice over 3dNow! or even x87 code for matrix multiplication. That 3dNow! outperforms SSE on the AMD CPU is interesting since the 3dNow! code is register deficient and the B matrix must be reloaded into memory for each row of the A matrix. It is fair to surmise that the AMD cache implementation keeps this reloading from becoming a significant performance killer, at least in the case of the 4x4 matrix multiplication.

Finally, the 3dNow! performance is compared to that of SSE for demo *Par03* in Figure 12.5c. In this demo (the Simpson's Rule numerical integration), the larger bit

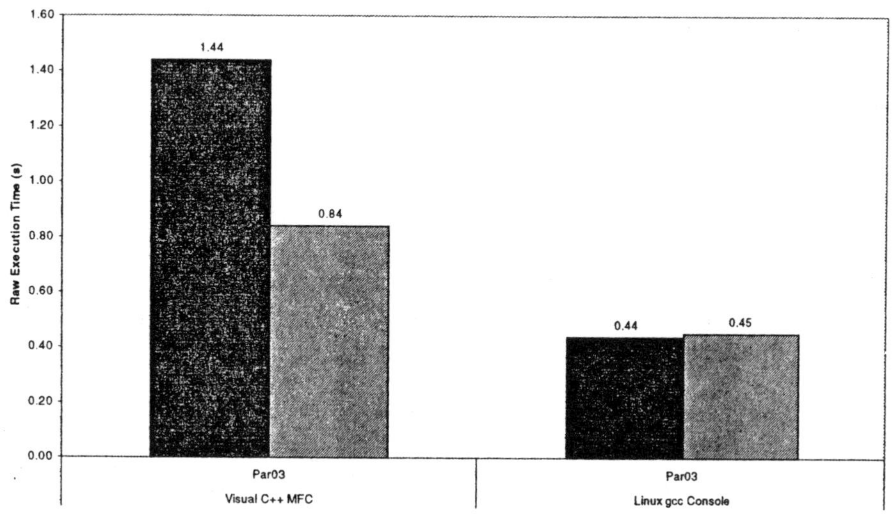

Figure 12.4: Demo *Par03* raw execution times on the test computer for 50,000 iterations of the Simpson's Rule Integration of x^5 over 10,000 subintervals. Compared are C compiled to x87 instructions and inline ASSEMBLY 3dNow! SIMD instructions.

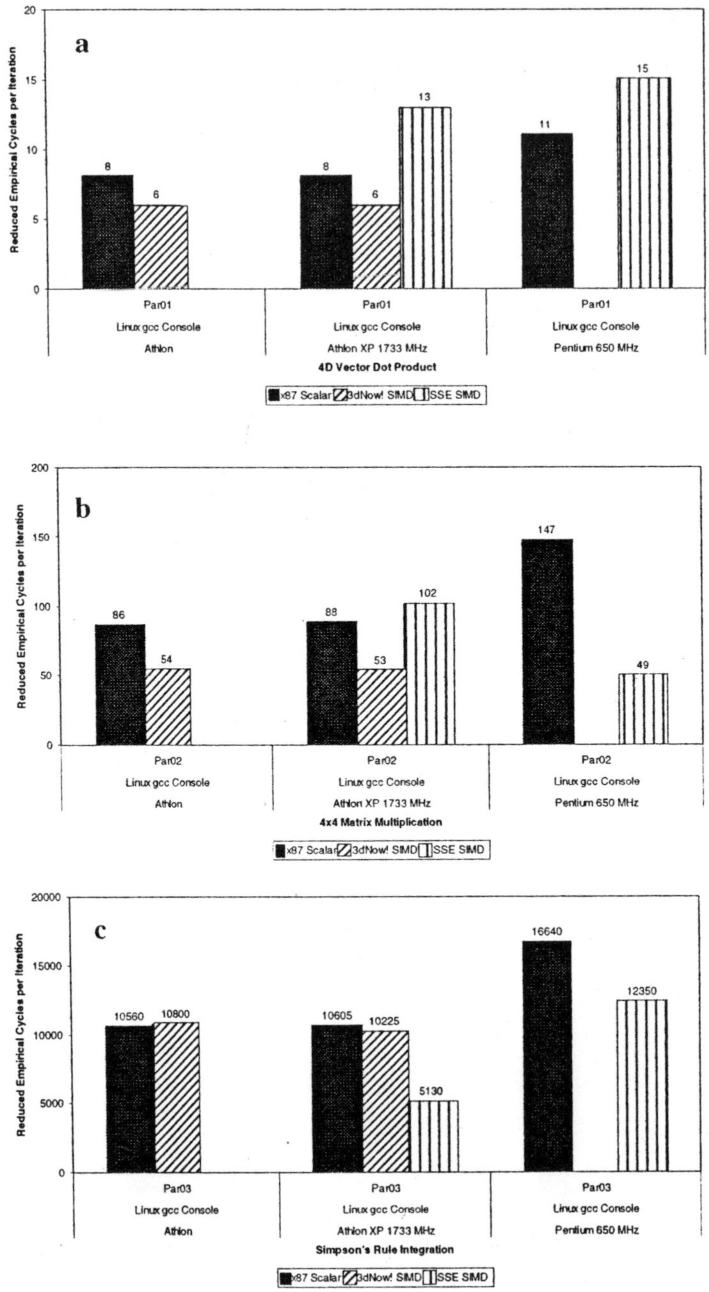

Figure 12.5: Reduced empirical cycles per iteration for Demos *Par01* – *Par03* for an Athlon 1200 (x87 and 3dNow! capable, adapted from Figures 12.1, 12.3 and 12.4), an AMD Athlon XP 2100+ (x87 and both 3dNow! and SSE capable) and a Pentium III (x87 and SSE capable). The SSE routine is *slower* than the x87 and 3dNow! routine for *Par01* on both AMD and Intel CPU's. For *Par02*, the Intel implementation of SSE is superior to x87 code, but the AMD implementation of SSE runs slower than both x87 and 3dNow!. SSE shines for *Par03* on both AMD and Intel CPU's.

width of the SSE instructions leads to a significant advantage on both Intel and AMD processors. In a sense, this increases an effective bandwidth of data to the processor. SSE does not suffer the performance hit as mentioned for the Dot Product since the ratio of the number of packed floating-point operations to the number of shuffles/scalar adds is much larger.

These comparisons again illustrate a key point mentioned throughout this text: general, highly portable code is not as fast as it possibly can be for the general architecture/platform. In addition, no specific processor or instruction set is the best for *all* situations. The Palomino and later Athlons providing multiple technologies have a distinct advantage that 3dNow! can be utilized for dot products and SSE for Simpson's Rule integration.

12.2.7 Compilers for SIMD and Additional SIMD Extensions

One limitation to using the extended x86 instructions is that most current compilers cannot compile to these op-codes. For example, MS VC++ 6.0 cannot, so inline ASSEMBLY must be used. To even use the inline ASSEMBLY with Visual C++, the 'processor pack' upgrade is needed so that the compiler recognizes the SIMD inline ASSEMBLY Language mnemonics. Intel's C++ and Fortran, gcc (the GNU compiler distributed with Linux) and VectorC (from CodePlay) are compilers that *can* produce SIMD instructions directly from source (that is, without the need for inline ASSEMBLY Language). However, gcc achieves this via the use of 'built-ins" that resemble inline ASSEMBLY Language. A demo version of VectorC for Windows is provided on the CD-Rom that accompanies this book. A number of useful MMX, 3DNow! and SSE instructions are summarized in Table 12.1.

The SIMD instruction set has been further extended to include the SSE2 instructions permitting double precision SIMD operations. There is also an extended 3DNow! instruction set. The Motorola/IBM processors utilize SIMD instructions for the AltiVec unit added in the MPC/7400 processor.

The SSE3 instruction set expands the SSE2 set. For example, SSE3 includes instructions for adding and multiplying the packed values within a single register, much like PFACC instruction in 3DNow!. This may reduce the scalar penalty the Intel instruction sets pay in the dot product and similar operations as discussed in section 12.2.6.

12.3 MIMD

For some algorithms, MIMD systems can achieve incredible performance compared to single processor machines. The analysis of an algorithm for MIMD execution can be complicated and should include system communication time, computation time and

Table 12.1: Some basic data transfer MMX and floating point 3dNow! and SSE SIMD instructions given in the Intel format. This is only a partial, sample list. hi refers to high order byte and lo is low order byte in a double word. hi 32 and lo 32 refer to upper and lower 32 bits in a larger data structure, etc.

Set	Instruction	Data	Operation
MMX	MOVD	32 bits	reg/mem$_{hi\|lo}$ <- 32 bit reg/mem
	MOVQ	64 bits	reg/mem <- 64 bit reg/mem
	PUNPCKDQ	32 bits	64 bit reg <- 32 bit reg/mem$_{lo\ 32}$
	PUNPCKHQ	32 bits	64 bit reg <- 32 bit reg/mem$_{hi\ 32}$
3DNow!	PFADD reg1, reg2/mem	2 Packed 32 bit floats	reg1$_{hi\|lo}$ <- reg1$_{hi\|lo}$ + reg2/mem$_{hi\|lo}$
	PFACC reg1, reg2/mem	2 Packed 32 bit floats	reg1$_{lo}$ <- reg1$_{lo}$ + reg1$_{hi}$ reg1$_{hi}$ <- reg2/mem$_{lo}$ + reg2/mem$_{hi}$
	PFMAX reg1, reg2/mem	2 Packed 32 bit floats	reg1$_{lo}$ <- max(reg1$_{lo}$, reg2/mem$_{lo}$) reg1$_{hi}$ <- max(reg1$_{hi}$, reg2/mem$_{hi}$)
	PFMUL reg1, reg2/mem	2 Packed 32 bit floats	reg1$_{hi\|lo}$ <- reg1$_{hi\|lo}$ * reg2/mem$_{hi\|lo}$
	PFRCP reg1, reg2/mem	2 Packed 32 bit floats	approx. reg1$_{hi\|lo}$ <- 1/(reg2/mem$_{hi\|lo}$)[a]
	PFRSQRT	2 Packed 32 bit floats	approx. reg1$_{hi\|lo}$ <- 1/sqr(reg2/mem$_{hi\|lo}$)[b]
	PREFETCH	8 32 bit floats	prefetch data into primary cache
SSE	MOVAPS reg1/mem, reg2/mem	128 bits	128 bit reg1/mem <- reg/mem
	MOVHPS reg1, reg2/mem	64 bits	reg1$_{hi\ 64}$ <- reg2/mem$_{hi\ 64}$
	MOVHLPS reg1, reg2/mem	64 bits	reg1$_{lo\ 64}$ <- reg2/mem$_{hi\ 64}$
	MOVSS reg1, reg2/mem	32 bits	reg1$_{lo\ 32}$ <- reg2/mem$_{lo\ 32}$
	ADDSS reg1, reg2/mem	32 bit float	reg1$_{lo\ 32}$ <- reg1$_{lo\ 32}$ + reg2/mem$_{lo\ 32}$
	ADDPS reg1, reg2/mem	4 Packed 32 bit floats	Packed reg1 <- reg1 + reg2/mem
	MULSS reg1, reg2/mem	32 bit float	reg1$_{lo\ 32}$ <- reg1$_{lo\ 32}$ * reg2/mem$_{lo\ 32}$
	MULPS reg1, reg2/mem	4 Packed 32 bit floats	Packed reg1 <- reg1 * reg2/mem
	DIVSS reg1, reg2/mem	32 bit float	reg1$_{lo\ 32}$ <- reg1$_{lo\ 32}$ / reg2/mem$_{lo\ 32}$
	DIVPS reg1, reg2/mem	4 Packed 32 bit floats	Packed reg1 <- reg1 / reg2/mem
	SHUFPS reg1, reg2/mem, imm8	4 Packed 32 bit floats	Shuffle 32 bit operand ordering; imm8 is an 8 bit flag that specifies where data from reg1 goes into which position in reg2[c]
	SQRTSS		
	SQRTPS		
	MAXSS		
	MAXPS		
	PREFETCH		prefetch data into primary cache

a. first approximation; to refine, use PFRCPIT1 and PFRCPIT2.

b. first approximation; to refine, use PFRSQIT1

c. See *IA-32 Intel Architecture Software Developer's Manual, Volume 2: Instruction Set Reference, Chapter 3* for a description of the bit encoding.

processor idle time. One algorithm that has been shown to exhibit considerable performance improvement in MIMD applications is matrix multiplication. The scalar versions shown in Ch. 7 and the SIMD version in Section 12.2 scale as n^3 for nxn matrices. Foster has presented several parallel matrix multiplication algorithms, one of which scales roughly as n^2/p, where p is the number of processes. Though this is clearly a better scalability, the algorithm does introduce additional overhead to the problem in form of data management. This is consistent with the idea presented throughout this book that higher performance code requires data management that is not always 'intuitive.'

As mentioned in Section 12.1.4, MIMD constitutes the most generally applicable approach to parallel processing. Multiprocessor machines generally fit the MIMD model. In addition, larger, more complicated architectures based on network communications allow very large numbers of processors (as well as memory and disk spaces) to be applied to a computational problem. Though highly specialized and research MIMD systems exist, three broad classes of MIMD computational systems are discussed here to pique the interest of the reader (that is, three that are relatively easy to implement): Network of Workstations, clusters and distributed systems. Each of these may include SMP nodes or utilize SIMD capable processors, so tiers of parallelism exist.

12.3.1 Networks of Workstations

The Network of Workstations (NoW) is a network of computers that are typically used for tasks involving considerable user interaction. However, processes can run on the workstations, usually when otherwise idle, that allow them to function as a parallel machine. In other words, a large priority difference exists between user interactive tasks and purely computational tasks running in the background.

Many sites utilize a NoW during off-hours (such as nights and weekends) to perform CPU intensive tasks while leaving the full workstation resources available to the user during 'on-hours.' This is an extremely efficient use of existing equipment when full CPU power is not needed for computation "24-7." Custom parallel packages for the NoW architecture employing the programming techniques in this book make a powerful tool for relatively small computations that only need be performed occasionally.

12.3.2 Clusters

A cluster can be very similar to a NoW, but is generally distinguished by the fact that each computer in the network is a node dedicated to computation. Whereas each node in the NoW has keyboard, monitor (and perhaps other devices) and is generally used interactively by a human user, the compute nodes in a cluster are 'stripped' of peripherals (even keyboards and monitors). The nodes, therefore, cannot be typically accessed for interactive use. As a result, virtually 100% of each node's cycles are available for computation in a cluster.

Clusters vary in size, design and networking topology. Each such system has its merits, and no single cluster design is perfect for all applications. Indeed, large clusters are typically divided into subclusters with their own functional specialization.

12.3.3 Distributed Computing

The connotation of terms "NoW" and "cluster" generally evoke systems located at a single site. The Internet allows connectivity of literally millions of computers around the world. For extremely large computational projects, gaining access to this processing power is difficult to resist. The process of using such widespread resources is distributed computing.

One of the most popular distributed computing projects is SETI@Home. Nearing completion of its first phase, the landmark project has utilized millions of CPU's to accumulate *over 1 million* CPU *years* since 1999, and currently adds just under 2000 CPU *years* per day. The "system" achieves tens of TFlOPS in sustained computing.

The SETI@Home model is conceptually simple. Radio astronomy data from the Arecibo Observatory is broken into work units that must undergo extensive CPU intensive signal processing in the search for 'real' signals from space. Users, over 6 million of them, download these work units to be processed on their home (or work) computers. Typically, the software is run as a 'screen saver,' an approach that uses cycles otherwise wasted drawing meaningless graphics on-screen to perform a scientifically significant computational task. When the work unit is complete, the analyzed data is uploaded back the server and a new work unit is retrieved.

There are other very large computational projects using the general distributed computing model, and several are listed in Table 12.2. Indeed, this method serves as a model easily adaptable to many, possibly smaller, projects. One does not need to make data available to the general users on the Internet to take advantage of this approach. For example, employee home computers, NoW at branch offices and

Table 12.2: Some current distributed computing projects that welcome public participation.

Project Name	Brief Description	URL
folding@home	protein folding	http://foldingathome.standford.edu
Genome@home	gene encoding	http://gah.standford.edu
GIMPS	search for Mersenne prime numbers	http://www.mersenne.org
Screensaver Lifesaver	Cancer protein-drug interactions	http://www.chem.ox.ac.uk/curecancer.html
Seti@home	search for extraterrestrial intelligence	http://setiathome.ssl.berkeley.edu

educational computer labs may be available to 'share cycles.' In addition, specialized compute centers are accessible online at which compute cycles may be purchased.

12.4 IMPLEMENTING MIMD

How is MIMD implemented? A common technique is the 'message passing' model in which nodes communicate status and synchronization information during the parallel execution. Messages are not always necessary during execution (that is, after the remote processes are created), especially when the parallel job consists of completely independent processes. After a few conceptual examples, the 'embarrassingly parallel' case is illustrated first. Also, none of the networking based examples in this section have *any* security based measures, so use them on open networks at your own risk.

The distribution of problems such as vector dot product, matrix multiplication and linear system solvers (such as Conjugate Gradient, Gauss-Seidel, etc) is not unique. For example, since the dot product, c = **a** + **b**, is

$$c = \sum_{i=1}^{n} a_i b_i$$

the system can be distributed over n processes. Each process is given one data pair, a_i and b_i, and returns the product. If the data must explicitly be sent to, and returned from, the working processes, there are 3n communications required to distribute (and return) the data. It is clear in this case that communication cost would likely outweigh the benefit of distribution. A shared memory approach in this instance would likely be better, so that each working process can act on the data directly without the 3n communications.

Other algorithms, such as matrix multiplication, can be completely redesigned to benefit from distribution. Other examples include finite differences, pair energy calculations, and optimization problems. These reworked algorithms may not resemble the 'serial' versions (that appear as the math on paper), a concept which again emphasizes a limitation to the Scientist Style as discussed throughout Part II.

12.4.1 Simple Parallel Computing without Messages

Sometimes a compute job contains such inherently concurrent components that the processes in the job are completely independent; these are often termed 'embarrassingly parallel.' Examples include computer-generated scenes in a motion picture, multi-determinant calculations in quantum chemistry problems, and Monte Carlo simulation of the same system at several temperatures. Indeed, any parameter study fits this category! The Monte Carlo procedure at various temperatures will serve as a test sample to illustrate such a calculation.

The sample computation uses an executable program similar to the Monte Carlo procedure as developed throughout this book. The key difference is a temperature loop wrapper around the general code that allows the simulation to be done at various temperatures. Further, the software needs to run in the background (to be called as needed), so it should get various inputs from an input file rather than via any direct user interaction. Similarly, output should go to an output file rather than to a screen window.

One could manually start each such process on each compute node. On a NoW, this would involve physically moving to each workstation to run the process. Clearly, this is a low efficiency approach. An alternative is to utilize scripts or remote process tools to activate processes on network resources, though this may still be quite involved. Obviously, when using more robust operation systems (Linux or Windows 2000 Server), there are more options in how to implement remote process management. However, to illustrate a straightforward method, Windows 98 (which is generally quite restrictive in regard to remote process management) will be used.

Parallel computing fits quite well with the client/server networking model, so perhaps it is best to use this approach. A simple solution involves the compute process to run as a daemon (in Unix/Linux) or service (in MS Windows) that 'listens' on the network for 'assignments' or other commands. This can be accomplished via an http server with computational tasks initiated via a browser for simple jobs. Similar to the manual approach mentioned above, this approach might also be quite tedious since processes are 'spawned' manually (though from one computer 'terminal'). It is possible to automate the spawning processes, but such a simplistic network paradigm lacks true message passing robustness. Other, more sophisticated, networking arrangements are likely more efficient.

Windows 98 (and other versions of Windows are similar) includes two tools that make implementing remote processes possible: Personal Web Server (PWS) and Web Based Enterprise Management/Windows Management Instrumentation (WBEM/WMI). The first, the http server, provides a 'way in' to the Win98 computer from the network. By executing a server side script (such as an Active Server Page) that creates a process, the remote user can run a program in the remote Windows 98 process space.

The process creation is done in two steps. The first uses the WMI moniker "winmgmts" in a call to the `GetObject` function of WMI with a script line such as

```
set process =
GetObject("winmgmts:{impersonationLevel=impersonate}
!<remote_computer_name>\root\cimv2:Win32_Process")
```

where <remote_computer_name> is the name of the computer on the network. The process object contains a creation function, and once this process object is created, a call can be made to the `process.create` function passing the name of the executable file name (and the process id number is returned):

```
result = process.Create (file,null,null,processid)
```

With these lines in a 'web page' script (such as an Active Server Page), a remote user can run a process on the server machine via an http browser. A "parallel computer" is thus created if each node in the system is running the web server with appropriate script pages, and the user submits the job to each node in turn.

This method of parallel implementation does not require any 'advanced programming' at the network or socket level. All that is needed is a web server and server-side scripting infrastructure that can be used to create processes. PERL, VBScript and PHP are examples of such scripting languages. Note that this is *not* an implementation running the simulation code in the scripting language (which would be very, very slow); rather, the scripting language is used merely as a bridge between the http server and the process management API of the operating system.

Of course, one is not limited to using an http server and server scripting to utilize networked computers as a parallel machine. Simple client/server applications can easily be coded using sockets. Sockets are OS objects that provide a common interface for network programming at a higher level. To create a server, a socket object is created and programmed to 'listen' on a given port. The server also grants a connection when an appropriate request is made by a client. Event handlers are written to process data when received. The client program only differs in that rather than listening to a port, connection requests are made to the server. Such an approach can be used to configure multiple computers when the parallel algorithm has data dependencies and requires message passing.

12.4.2 Simple Master-Slave Parallel Computing

Useful in networked parallel computing is the 'master/slave' concept. Slave nodes are pure compute nodes; they perform little or no administrative function. The master node, however, controls the parceling of work units to the slave nodes as well as reconstructing the final solution when all computation is complete. In addition, in some parallel systems, the master node will mediate all internode slave communications (when the slaves do not communicate with each other directly). Not all parallel systems require the master/slave arrangement, but it is quite useful when parallel computing resources are not intrinsic in the OS kernel.

Though it may seem confusing, in this illustrative example, the slaves are programmed as "servers" listening for connection requests on a port. The master node is acting as a client, making the request. This approach allows the parallel system to be specified from the master node, whereas the opposite configuration would require each slave node to request connection to the master. In other words, a list of available slave node ip addresses (or machine names, if a name server is used) is maintained on the master with connections originating from the master. After each slave is connected, the master node supplies appropriate data to each slave process. In a graphics example, four slaves can compute four scenes simultaneously, with no inter-process communication needed.

The generic program design for this paradigm is shown in Figure 12.6. Note how easily this approach can be adapted to any parallel computation requiring no inter-process communication.

12.4.3 Simple Message Passing Parallel Programming

Many parallel computing problems involve considerable data dependencies or other inter-process communication requirements. For example, using the Alternating

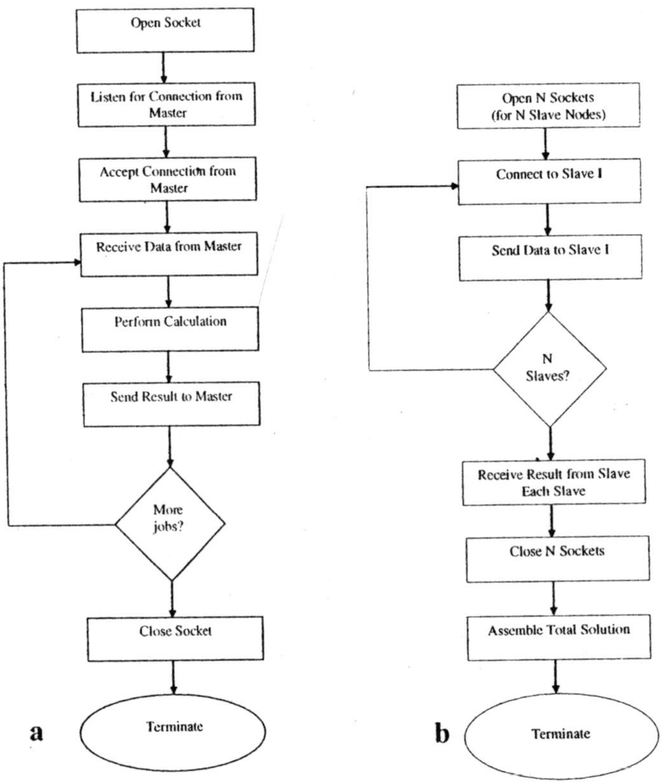

Figure 12.6: Basic procedure for coding parallel programs using socket objects. (a) After process creation, each worker process listens on the socket for data and instructions from the server process. (b) The server process sends data to slaves and waits for results.

Direction Implicit (ADI) method, a Laplace equation solver in two dimensions can be divided into work units, but between iterations, synchronization of the full dataset is required. Similar techniques are used to solve the Navier–Stokes flow field equations in the Finite Volume method.

To accomplish the data synchronization, information is passed between the slave nodes and master node or even between the slave nodes directly. To illustrate the technique, the Laplace solver is considered. Without going into mathematical detail, a solution *u(x,y)* is sought to the partial differential equation

$$\nabla^2 u(x, y) = \frac{\partial^2 u}{\partial x^2} + \frac{\partial^2 u}{\partial y^2} = 0$$

subject to the Dirichlet boundary conditions (i.e., u is prescribed on the boundary) of the given problem. If the computational domain is divided into an n x n mesh, the general derivation of a numerical solution to this equation results in a system of n linear equations in n unknowns (that is, the n unknowns are the values of u at each mesh point). For each mesh point, the solution is

$$u(x, y) = 1/4 * [u(x+h, y) + u(x, y+h) + u(x-h, y) + u(x, y-h)]$$

where h is the mesh size. Grouping of common *u(x,y)* results in a system of equations (for the i,j^{th} mesh point). Applying the ADI method (which tridiagonalizes the matrix of these n linear equations) divides the problem into two 'components': one that couples the 'rows' of unknowns to each other, but eliminates the column coupling, and one that couples the 'columns' of unknowns to each other, but has no row-row coupling. In particular, there are n sets of n coupled equations (for each of the n rows), each of the form

$$u_{i-1,j}^{m+1} - 4u_{ij}^{m+1} + u_{i+1,j}^{m+1} = -u_{i,j-1}^{m} - u_{i,j+1}^{m}$$

where m is the iteration number. Likewise, for the n columns, the sets of equations can be written

$$u_{i,j-1}^{m+2} - 4u_{ij}^{m+2} + u_{i,j+1}^{m+2} = -u_{i-j,j+1}^{m+1} - u_{i+1,j}^{m+1}$$

A complete iteration involves solution of both components. This system of 2 * n sets of n equations must each be solved by some iterative procedure, such as the Gauss-Seidel Iteration.

Each set could be solved in parallel, considerably reducing the total execution time for the entire process. Consider a parallel system with n nodes. For each component (solving for $m+1$ using rows or $m+2$ using columns as described above), there are n sets of n coupled equations in n unknowns. The work unit for each node would therefore be a single set of n equations to be solved for n unknowns. This can be visualized by considering that for a 100 x 100 mesh, there are 10,000 unknowns. A system with 100 compute nodes could be used so each node solves for 100 unknowns, using data in the other rows or columns as "mini" boundaries. Once each node solves for its set of 100 unknowns, the results must be communicated to the other nodes, especially between iterations, m; otherwise the system is not properly coupled. Specifically, the subsets of 100 unknowns are 'recombined' into the entire set of $u(x,y)^{m+1}$ before the data is resubmitted to each node for computation of $u(x,y)^{m+2}$

12.5 MESSAGE PASSING TOOLS

The previous section outlined a 'raw' approach to parallel computation in which all communications were custom coded. For some high-demand applications, such custom coding may be the highest performance approach, but does require more coding effort. There exists many tools to aid in the generation of parallel code, and a popular and easy one to implement is the Parallel Virtual Machine (PVM); another is the Message Passing Interface, or MPI. However, the same general statements regarding the use of libraries that were made in Chapter 6 apply; there is performance overhead involved in using these systems.

12.5.1 Parallel Virtual Machine (PVM)

PVM is both a software suite used to create and administer NoW and cluster communications and a code library to compile parallel code. As a software package, PVM allows one to connect multiple computers (even running different OS's on different hardware) via a simple network; all the machines running PVM act as a single parallel computer with PVM function calls used to create processes and handle message passing. PVM inherently operates on the master-slave model. There are also several administrative tools for PVM. The web site for downloading PVM is given in Appendix A.

As a code library, PVM allows C (or Fortran) code to contain calls to the PVM software. Such calls include process management and message passing. Since the calls are made available as a library, PVM should be able to be used with any C or Fortran compiler that is platform compatible.

One of the merits of PVM, especially to beginning parallel programmers, is the relative ease with which parallel systems and applications can be constructed. The PVM system itself is easy to implement, allowing one to create a parallel computer

Table 12.3: Some web sites offering sample PVM code.

Web Site	Sample Given
www.csc.fi/programming/examples/pvm/node2.html	Vector Dot Product in C
www.csc.fi/programming/examples/pvm/node3.html	Laplace Solver in Fortran
www.csc.fi/programming/examples/pvm/node5.html	Transpose of a square matrix in Fortran
www.itec.uni-klu.ac.at/~harald/mm.html	Matrix Multiplication in C
www.netlib.org/pvm3/book/node51.html	Vector Dot Product in Fortran
www.netlib.org/pvm3/book/node55.html	Matrix Multiplication in C
www.netlib.org/pvm3/book/node57.html	Heat Diffusion Solver in C

of several nodes merely by adding nodes via the administration tools. Though the PVM library contains many functions, only about five functions are needed for basic parallel application development. It is reasonable for a moderately accomplished programmer to set up a parallel machine on an existing network and write a small parallel application in a single afternoon. Web sites offering sample code for PVM are listed in Table 12.3.

Though MPI programs may be faster, PVM has three key strengths that should be kept in mind. First, PVM can be implemented on heterogeneous networks, such as Beowulf clusters built from different architectures. In this regard, it could be said that PVM programs are very portable. Second, PVM is very fault tolerant. Programs can respond to system load changes, even to the degree of a node becoming unavailable. Last, PVM has well developed dynamic resource management tools and process control functions.

12.5.2 Message Passing Interface (MPI)

A more robust, though similar, system is the Message Passing Interface (MPI). MPI began its 'life' in the mid 1990's as a standard for implementing large multi-processor machines. At that time, vendors of commercial MPP's (Massively Parallel Processors) each had developed proprietary message passing protocols so that source code was not portable. To address this, the standard for message passing implementation was introduced. The MPI implementation now has over 200 functions. Implementations of the MPI standard include MPI-LAM (Local Area Multicomputer) and MPI-CH. Web sites for downloading these MPI implementations are given in Appendix A.

MPI implementations are generally faster on MPP computers. MPI contains more point-to-point communication options and even allows for the definition of logical network topologies. However, MPI originally did not include the level of fault tolerance as PVM, this being sacrificed for speed (keeping in mind that on an MPP, the computer is more of a single system as opposed to clusters of individual systems).

Like PVM, MPI contains user level message handlers in their libraries. Both the MPI-LAM and MPI-CH implementations support heterogeneous networks.

Interestingly, MPI implementations and PVM can co-exist on a given network of computers. This allows the greatest flexibility in terms of user/programmer preferences. In many cases, a programmer does not need to choose one or the other, but can utilize the API that best suits the needs of the problem.

12.4.3 Other Tools

A great number of other tools are available for parallel computing, and some are listed in Appendix A. Some of these involve actual kernel level modifications, such as MOSIX (the OpenMosix How-To is on the CD Rom). Even Windows has extensions that allow clustering. In addition, High Performance Fortran is a compiler that is used to 'automatically' compile code for parallel execution, especially for data parallel problems; however, as with other optimizations, allowing the compiler to 'make decisions' on code structures may not always produce the 'best' solution. This Chapter has just scratched the surface of parallel computing, and the interested reader is encouraged to consult books that focus on parallel computing.

Appendix A

A List of Modern Development Tools

This appendix contains a brief list of several modern programming tools. These tools generally fit the theme of the book overall in regard to high performance programming and/or code optimization. This list is not intended to be exhaustive; the absence of a tool from this list does not imply anything about the quality or relevance of the tool. Further, the author has not personally tried all the tools on this list, so of course no warranty is given as to the usefulness, stability or value of these tools.

Ada, List of Free Compilers

http://www.adahome.com/Resources/Compilers/Free.html

This site contains links to numerous free Ada compilers. Ada was initially designed with a special emphasis on stability.

Ada, List of Commercial Compilers

http://www.adahome.com/Resources/Compilers/Vendors.html

This site contains links to numerous commercial Ada compilers.

AMD Core Math Library

http://www.amd.com/us-en/Processors/DevelopWithAMD/0,,30_2252_2282,00.html

From the site: "The AMD Core Math Library (ACML) incorporates BLAS, LAPACK and FFT routines, which can be used by a wide range of software developers to obtain excellent performance from their applications running on AMD platforms. The highly optimized library contains numeric functions for mathematical, engineering, scientific and financial applications."

APPSPACK

http://software.sandia.gov/appspack/

From the site: "APPSPACK is a software package for an asynchronous parallel pattern search."

Alternatives to Numerical Recipes

http://math.jpl.nasa.gov/nr/nr-alt.html

This page contains lists of numerical libraries and programming resources.

C++ Builder Studio

http://www.borland.com/cbuilder/index.html

This page markets Borland's commercial C++ compiler.

Code Profiler (for C++)

http://www.codeproject.com/cpp/profiler.asp

This site is an article on profiling, and includes a free, downloadable profiler.

CodeWright

http://www.borland.com/codewright

This page markets Borland's code editor.

Computational Gasdynamics – Free and Low Cost CFD Software

http://capella.colorado.edu/~laney/software.htm

This site contains links to categorized lists of free and low cost CFD applications and libraries.

Delphi Studio

http://www.borland.com/delphi/index.html

This page markets Borland's Delphi compiler.

Dictionary of Programming Languages

http://cgibin.erols.com/ziring/cgi-bin/cep/cep.pl?_total=1&_format=full&_userlink=1

This is a catalog of many programming languages. This list includes descriptions, histories, sample code and is cross-referenced among similar languages.

DMOZ Open Directory Project Programming Language Page

http://dmoz.org/Computers/Programming/Languages/Compiled/

FFTW

http://www.fftw.org

From the site: "FFTW is a C subroutine library for computing the Discrete Fourier Transform (DFT) in one or more dimensions, of both real and complex data, and of arbitrary input size. ... [FFTW] is free software ... Our benchmarks, performed on on a variety of platforms, show that FFTW's performance is typically superior to that of other publicly available FFT software. Moreover, FFTW's performance is *portable*: the program will perform well on most architectures without modification."

Forth Compiler List at the Forth Interest Group Home Page

http://www.forth.org/compilers.html

This site contains links to Forth compilers.

The Fortran Store

http://www.fortran.com/fortran/home.html

This site contains links to commercial and free Fortran code and products.

Free Compiler Catalog

ftp://ftp.idiom.com/pub/compilers-list/free-compilers

From the site itself: "This list catalogues freely available software for language tools, which includes the following: compilers, compiler generators, interpreters, translators, important libraries, assemblers, etc. — things whose user interface is a language. Natural language processing tools may also be included."

Free Pascal Compiler

http://www.freepascal.org/

This free, downloadable Pascal compiler is available for numerous platforms, including Linux and Win32.

Free Scientific Software under Linux

http://esca.atomki.hu/dlug/tudapps/linux_old.html

This site includes a list of both commercial and free programs and tools, including libraries and GUI tools.

gcc

http://gcc.gnu.org/

GNU Compiler Collection included with Linux distributions. gcc compiles code for: Ada, C/C++, Fortran, Java and Objective C. One of the strengths of gcc is that it can produce architecture specific code for a wide variety of architectures.

Glow Code Profiler

http://www.glowcode.com/

This commercial profiler for Windows software written in C++ can help detect memory leaks, performance bottlenecks and provides performance related statistics to the programmer

GoASM Free Win32 Assembler

http://www.jorgon.freeserve.co.uk/

This site not only lists the free Assembler, but many free tools (debugger, text editer, etc) are available as well. The site also includes Beginner, Intermediate and Advanced Tutorials on Windows Programming and Win32 Assembly code samples.

Google Directory Programming Languages Page

http://directory.google.com/Top/Computers/Programming/Languages/

Help Systems for Windows Programs

http://www.smountain.com

This site has numerous articles and tips for adding both WinHelp and HTML Help to Visual Basic and Visual C++ programs in Windows.

High Performance Fortran

http://www.crpc.rice.edu/HPFF/

This is the 'High Performance Fortran home page.' The page contains links to compilers, references, projects, benchmarks, etc.

IBM Development Kits

http://www-106.ibm.com/developerworks/java/jdk/index.html

IBM's JAVA Development Kit available for download.

Interactive Data Language (IDL)

http://www.rsinc.com/idl/index.asp

IDL is a commercial language specifically designed to rapidly develop data acquisition and visualization applications. At last check by this author, IDL was available in two license formats: per application or unlimited. Both licenses were quite expensive.

Intel C++ for Linux

Software/products/compilers/clin.html

This page has info about Intel's commercial C++ compiler for Linux.

Intel C++ for Windows

Software/products/compilers/cwin.html

This page has info about Intel's commercial C++ compiler for Windows.

Intel Fortran for Linux

Software/products/compilers/flin.html

This page has info about Intel's commercial Fortran compiler for Linux

Intel Fortran for Windows

Software/products/compilers/fwin.html

This page has info about Intel's commercial Fortran compiler for Windows.

Introductory Java for Scientists and Engineers

http://www.jscieng.co.uk/

This is a marketing page for the book Introductory Java for Scientists and Engineers by Richard Davies.

ISML Math Library

www.vni.com/products/imsl/docs/FNLMathDescrip.html

This page has information regarding Visual Numerics ISML Math Library.

JAMA

http://math.nist.gov/javanumerics/jama/

From the site: "JAMA is a basic linear algebra package for Java. It provides user-level classes for constructing and manipulating real, dense matrices. It is meant to provide sufficient functionality for routine problems, packaged in a way that is natural and understandable to non-experts."

Jampack

ftp://thales.cs.umd.edu/pub/Jampack/Jampack/AboutJampack.html

"A Java Package for matrix computations."

Java JIT

http://wwws.sun.com/software/solaris/jit/

This page gives information, and a download link, for Sun's Just In Time compiler for Java.

Java Number Cruncher

http://www.softpro.com/0-13-046041-9.html

This page markets the book Java Number Cruncher: The Java Programmer's Guide to Numerical Computing by Ronald Mak.

Java Numerical Toolkit

http://math.nist.gov/jnt/

From the 'Prospectus' page: "In this project, we seek to improve the state of scientific computing in Java by ... evaluating the suitability of the Java environment for scientific computing applications in order to gain the expertise to advise scientific software developers on how best use of its features". The site also includes preliminary performance results for this toolkit.

JavaNumerics

http://129.6.13.90/javanumerics/

From the site: "The JavaNumerics page provides a focal point for information on numerical computing in Java."

JMSL

http://www.vni.com/products/imsl/jmsl.html

This page markets Visual Numerics commercial library for Java.

jProbe

http://www.klgroup.com/jprobe/profiler/

This page markets a commercial Java code profiler.

LAM-MPI (Local Area Multicomputer – MPI)

http://www.lam-mpi.org/

LAM-MPI is a heterogeneous cluster implementation of MPI. It can be used for dedicated clusters or Networks of Workstations. At the time of this writing, the stable release is Version 6.5.9 and is available for download from the listed link.

libSIMD

http://libsimd.sourceforge.net

From the site: "The main goal is to provide scalar, vector, matrix, trigonometric, complex number, quaternion and FFT operations in the form of a dynamic library which can be compiled with GCC." This project is currently under development but has promise to grow into a very useful library utilizing the 3dNow!, SSE and SSE2 x86 instruction set extensions. Contributors are welcome to contact the developer.

Linux Assembly

http://linuxassembly.org

This page is for Assembly Language programming on Linux, for 'all' platforms. Much of the site's copy refers to Unix rather than Linux, suggesting the material is easily applicable to other Unix variants.

Linux Documentation Project

http://www.tldp.org

This is the home page for The Linux Documentation Project, which is a collection of "How-To" documents ranging from the very general to the very specific. Included are How-To's for gcc, Assembly Language, parallel processing and other topics of interest to programmers.

Linux Online – Development Libraries

http://www.linux.org/apps/all/Development_tools/Libraries.html

Linux Online – Development Languages

http://www.linux.org/apps/all/Development_tools/Languages.html

Linux Online – Development Tools

http://www.linux.org/apps/all/Development_tools/Tools.html

Make Java Fast: Optimize

http://www.javaworld.com/javaworld/jw-04-1997/jw-04-optimize.html

This site is an article by Doug Bell. Though a Java oriented article, Bell outlines several general points about optimizing code that are applicable to any language.

Making Numerical Java Programs Execute Faster

http://www.cs.uiuc.edu/Dienst/UI/2.0/Describe/ncstrl.uiuc_cs/UIUCDCS-R-2002-2310

This site is the Java optimization article by Zehra Noman Sura. From the site: " [Its language] features can cause Java to be significantly less efficient than FORTRAN or native C/C++ implementations. Since performance is critical in the domain of numerical computing, the use of Java for numerical programs can be limited. We present a technique to make numerical Java programs run faster."

MASM32

http://www.movsd.com/masm.htm

This Win32 Assembler is designed for experienced programmers already somewhat familiar with the Win API. Additional tools are available to extend the development environment.

Microsoft Software Development Kit (SDK)

http://msdn.microsoft.com

The SDK is designed to help develop Win32 code using the Win API. The file download is free but quite large.

MinGW

http://www.mingw.org

MinGW is a gcc variant that includes libraries for producing Windows code. At the time of this writing, MinGW supported C, C++, Objective-C and Fortran77, with ADA and Pascal available.

Miracle C

http://www.c-compiler.com/

Miracle C is a shareware (free to download) C compiler for Windows. This is an excellent learning tool for beginning programmers, since warnings and errors are not reported. This forces the programmer to study the code when a program does not run as expected.

mpC

http://www.ispras.ru/~mpc/

From the site: "mpC is a high-level parallel language (an extension of ANSI C), designed specially to develop portable adaptable applications for heterogeneous networks of computers."

Monte Carlo Software and Numerical Tools

http://www.cooper.edu/engineering/chemechem/MMC/software.html

This site contains links to software tools for the "Monte Carlo Practitioner."

MPI-CH

http://www-unix.mcs.anl.gov/mpi/mpich/indexold.html

MPI-CH is a free parallel machine implementation of MPI. Versions are available for 'Unixes' and Windows NT/2000.

MS Processor Pack

http://msdn.microsoft.com/vstudio/downloads/tools/ppack/

The C++ processor pack is needed for VC++ 6.0 to compile SIMD inline ASSEMBLY instructions.

NetLib

http://www.netlib.org/

From the site: "Netlib is a collection of mathematical software, papers, and databases." The collection includes algorithms from ACM Transactions on Mathematical Software and other peer reviewed sources.

NESL

http://www-2.cs.cmu.edu/~scandal/nesl.html

NESL is a high performance language for developing parallel programs. At the time of this writing, NESL is formally available only for 'big iron' Unix workstation machines, such as Cray C90 and IBM SP-2. Linked from this site is a comparison of NESL to an MPI based program, as well as number papers about NESL.

OpenMP

http://www.openmp.org

From the site: "The OpenMP Application Program Interface (API) supports multi-platform shared-memory parallel programming in C/C++ and Fortran on all architectures, including Unix platforms and Windows NT platforms."

Optimizeit

http://www.borland.com/opt_profiler/index.html

Borland's code profiler for the MS .Net Framework

OR-Objects

http://opsresearch.com/OR-Objects/index.html

This page markets a commercial Java library. From the site: "OR-Objects is a collection of 500 Java classes for developing Operations Research, Scientific and Engineering applications. It contains data structures and algorithms for developing problem specific solutions as well as implementations of classical algorithms."

Programming Languages

http://www.cs.waikato.ac.nz/~marku/languages.html

This site is a categorized list of programming languages.

Parallel Virtual Machine (PVM)

http://www.csm.ornl.gov/pvm/pvm_home.html

From the site: "PVM (Parallel Virtual Machine) is a software package that permits a heterogeneous collection of Unix and/or Windows computers hooked together by a network to be used as a single large parallel computer."

PVM with MFC

http://mywebpages.comcast.net/brian.piscopo/pvm-web.htm

This site outlines tips to get PVM working with the Microsoft Foundation Classes (MFC) and the VC++ multithreaded library.

Python

http://www.python.org

From the site: "Python is an *interpreted, interactive, object-oriented* programming language. It is often compared to Tcl, Perl, Scheme or Java."

REDUCE

http://www.uni-koeln.de/REDUCE/

From the site: "REDUCE is an interactive program designed for general algebraic computations of interest to mathematicians, scientists and engineers."

S

http://netlib.bell-labs.com/cm/ms/departments/sia/project/S/index.html

S was designed in the statistics research group at Bell Labs. From the site: "[S] provides rapid high-level prototyping for computations with data, featuring interaction, graphics, and universal, self-describing objects.

ScaLAPACK

http://www.netlib.org/scalapack/

ScaLAPACK is a collection of Linear Algebra libraries for use with MPI.

Scientific Applications on Linux (SAL)

http://sal.kachinatech.com/sal1.shtml

SAL is a categorized list of Linux applications, tools, etc. of particular interest to scientists and mathematicians. Since many of the products are "Open Source," it should be possible to port to other platforms.

SciLab

http://www-rocq.inria.fr/scilab/

SciLab is an open scientific computing package for use on Linux and Win32. A parallel version is implemented via PVM.

SciMath

http://www.scimath.com/

SciMath is a commercial C/C++ scientific math library.

shuJIT

http://www.shudo.net/jit/

This page offers information on the shuJIT downloadable Java Just In Time Compiler.

Sourceforge – Scientific Application Projects

http://sourceforge.net/softwaremap/trove_list.php?form_cat=97

This Sourceforge list contains open source collaborative projects related to scientific applications. Programmers may find useful applications, libraries, etc. in this list. In addition, programmers may contribute to the development of these projects. Programmers are generally encouraged to contribute to existing projects rather than starting new ones.

Sourceforge – Software Develoment Projects

http://sourceforge.net/softwaremap/trove_list.php?form_cat=45

This Sourceforge list contains open source collaborative projects related to the software development tools. Programmers may find useful libraries, editors, profilers, etc. in this list. In addition, programmers may contribute to the development of these projects. Programmers are generally encouraged to contribute to existing projects rather than starting new ones.

Thomas Tridiagonal Matrix Algorithm (TDMA)

http://www.aoe.vt.edu/~mason/Mason_f/CAtxtAppH.html

This page includes a link for a Fortran TDMA solver as well as a pdf manual for using the code.

Top Ten Numerical Algorithms

http://www.computer.org/cise/articles/Top_Algorithms.htm

This is the first page of Jack Dongarra's guest editorial to *Computers in Science and Engineering*. There are descriptions and articles about the mentioned algorithms.

VectorC

http://www.codeplay.com

VectorC is a C compiler designed for producing high performance code. VectorC produces MMX, SSE and 3dNow! SIMD instructions from native C source code, eliminating the need for the programmer to use inline Assembly to get SIMD optimizations on the x86 architecture. At the time of this writing, VectorC was available for Windows; a Linux version is under development. A demo version of VectorC for Windows is provided on the cd.

VHDL

http://www.vhdl.org/

VHSIC Hardware Description Language tools and components. This language focuses on the design of digital integration circuits.

Visual C++ 6.0

http://www.discount-software.ws/microsoft/visual-c-plus-plus-6.0-pro.html

For those who wish to avoid the newer MS VC++ .Net compiler, this is one source for the VC++ 6.0 Pro compiler; may be out of stock.

Visual C++ .Net

http://msdn.microsoft.com/visualc/

This is the product marketing page for the MS Visual C++ .Net. The marketing information emphasizes web and network development.

Vtune

http://developer.intel.com/software/products/vtune/index.htm

Vtune is Intel's commercial code performance analyzer (profiler). An online demo is provided at the site.

Appendix B:

Buying a High Performance PC

"You get what you pay for." This colloquialism applies to computer purchases as well as anything else. Too often, computers priced significantly lower than competitors demonstrate why when taxed on a serious computational problem. Reliability is a related issue; cheaper systems, composed of cheaper components, tend to wear faster and fail more often. This appendix outlines some of things to consider when purchasing a PC for high performance computing. No specific brand name is endorsed, but (non-exhaustive) lists of companies currently offering performance components are given. Each subsystem of the computer will be considered in turn. Again, the gaming community serves as guide to performance hardware, so a search of hard-core gaming web sites provides the reader with current, performance PC hardware.

This outline applies specifically to x86 based PC systems. In general, however, the suggestions can serve as guidelines to apply to any architecture; there simply does not exist the space in this book to offer detailed comparisons of *all* possible hardware.

B.1 BUILDING VS. BUYING

This back-and-forth question is often seen in computer books. Buying 'off-the-shelf' computers, such as those sold in department stores, is fine for Internet surfing, word processing or maintaining a list of family videos in a database, but these systems are not designed with true high performance in mind. Further, not all systems of a given CPU clock speed are the same. In side-by-side comparisons, using real scientific software (Quantum Chemistry computations), the author's custom built 1200 MHz Athlon based system outperformed a similar system *using the exact same CPU, hard drive and other peripheral devices!* That is, the physical CPU, RAM, hard drives and other components were removed and placed in different main boards and the performance measured. The differences were astonishing.

The key problem with mass produced department-store computers is that they are assembled to appeal to the mass consumer. That is, they are designed around features believed desirous of a *general* computer user, and priced accordingly. So, who is the general computer user? In today's computer market, based on the capabilities of the mass produced systems, the general user surfs the Internet (ie, uses the computer as a browser terminal), views/manipulates small digital images, runs children's software, plays casual (non-high performance) games, etc. While systems may be advertised with competitive, state-of-the-art CPU clock speeds, other system components are of generally low quality. Generally, home users have stated, through purchasing trends, they do not want to buy capability they don't use.

The key distinction to be made is the computer built for 'home use' versus one designed for 'professional use.' Business grade computers are more expensive than home grade computers because they are built from higher quality components. In the pure business environment, reliability is the principle factor: businesses require systems with low failure rates. The idea extends equally well to other professional 'classes,' specifically high computational performance. It is reasonable, then, that high performance computers cost more (though not that much more, as will be discussed below).

Some mass production houses (and the department stores they supply) do build 'custom' computers, but a word of caution is warranted. In this author's experience, these companies just cannot break their mass-production habits. Specifically, a group of research computers once ordered from a popular computer company (one that is still in business and advertises extensively on television) arrived with missing, incorrect or malfunctioning hardware. I got the impression that the staff assembling the system just did not believe I really wanted the system configured as I ordered! To exasperate these problems, we were billed for hardware not purchased, and it literally took many months to settle the issues of what was actually purchased vs. what was shipped. Anecdotally, the author has heard similar stories about this same company from other system administrators. In defense of this company, however, I have also heard home-user class customers report being quite pleased with the systems, but that's the point. They are 'home user' class systems not particularly suitable for 'professional' use.

Building a custom computer with performance in mind is not as difficult as it seems. For those that enjoy reading reams of technical specifications on different hardware, selecting 'that right match' of hardware can be done oneself. On the other hand, many small, professional-owned computer stores specialize in custom building computers to meet the user's needs. This includes high performance. As additional side benefits, these smaller stores tend to provide much better equipment, better after-sale service and tend to support the local economy in the buyer's own area. The owners and staffs of these stores are mostly persons who take a serious interest in computer equipment; this is in direct contrast to the department stores whose staff is more likely a broad mix of people with perhaps no clear interest in the merchandise.

The rest of this Appendix addresses the issues to be considered when custom building a computer, whether you are doing it or having it built. If you are having a computer built, these guidelines can assist you in providing questions to ask the actual builder or to supply the builder with information needed to build the system you desire.

B.2 THE STARTING POINT IN CUSTOM BUILDING

The first step in buying a computer is to define the *use* of the system. As mentioned in the previous section, computer hardware appropriate for casual home use is quite different from that required for serious computing. Even 'serious' users can be subdivided. In general, CPU intensive tasks fall into one of two main categories: floating point intensive tasks and memory access/transfer intensive tasks. Graphical applications, such as video image processing, are very memory intensive. On the other hand, simulations, such as quantum mechanical calculations, are floating point math intensive. They may overlap somewhat, but applications are typically either more memory or math intensive.

The applications planned will determine the OS needed, which in turn may influence the hardware required. If you are developing software from scratch, you can choose an OS (and thus the hardware) based on preferred features. Considerations include portability (for example, Windows applications are generally *not* very portable, either to other platforms or sometimes to different versions of Windows), OS function availability and deployment/distribution opportunities (if developing a code to be distributed).

How does the OS choice influence the hardware? One way is driver availability. MS Windows has evolved the Plug N Play concept to the point that many hardware manufacturers save money by excluding key hardware from the boards; the 'missing' functionality is emulated in software. Note that this is a key way that 'home user' systems are cheaper. It is also one way the performance of such systems can suffer: hardware is generally faster than software. A system in which code execution by the CPU must 'compensate' for missing hardware suffers two ways. The peripheral itself is slower than it could be and CPU cycles are expended doing work that could be done in peripheral hardware.

A classic example of this is internal modems. Inexpensive Plug N Play modems purchased to work with MS Windows often lack the on-board controller; its functionality is supplied by the software driver. These modems, in essence Windows driven modems, are often called "Winmodems," and they generally will not operate in Linux (some software drivers have been written for such modems to operate in Linux, making this subset the so-called "Linmodems"). Hardware modems, on the other hand, possess the on-board (on the modem card itself) controller and require no software driver. Therefore, hardware modems are OS independent. Other examples of this include network interface cards (NICs) and printers.

Obviously, therefore, the system components are not independent. In addition to the OS interacting with the system, the individual components interact as well. This may seem like an obvious statement, but it is mentioned since it can easily be overlooked when designing a system. One may find 'that perfect main board' only to find that it does not support the chosen CPU. Therefore, it is important to design a system in a logical manner, starting with the components that most influence possible capability of the others. At each stage of the design process, cost is weighed vs. desired performance.

Before discussing the major system components, a few words about benchmarking may be useful. Hardware selection should be made based on actual performance data in *real* applications as much as possible. It is recommended that artificial benchmarking tests be viewed with skepticism, or at least viewed with acceptance of their questionable value. Further, the Internet is flooded with benchmarking tests done using non-scientific testing methods (for example, a "CPU comparison" may well involve systems with different main boards, memory, main board configurations, software optimizations, etc); there are many people running canned benchmark programs without the technical skills to properly evaluate the systems or the data produced. A general rule of thumb the author follows is to rely on benchmarking data that includes reasonable technical descriptions of the data, including *why* the test resulted in one system outperforming another.

A corollary to the above discussion is the danger of misinterpretation of benchmark data produced from programs compiled with old or biased compilers. For example, a compiler that 'favors' production of the "3DNow" instructions for AMD CPU's will generate a benchmark that favors AMD chips, whereas code compiled with SSE2 instructions will favor Pentium4 CPU's. Many benchmarks claim to be the definitive or neutral test, but bias may still exist. The best bet is to seek benchmarks performed with *real* applications. If you are custom coding an application, it will be a bit harder to evaluate the test data to determine which hardware is more suitable for a given task.

B.3 CPU

A general rule for CPU selection is to purchase the highest clock speed one can afford. However, the clock speed itself is not the definitive CPU specification; indeed, recent advances in PC CPU production have blurred valid comparison between different manufacturers. Further, chips with the same clock may have different cache sizes, different floating-point processors, different memory space accessibility, different configuration capability, different pipeline specifications, different branch prediction capabilities and different advanced instruction sets.

What is meant by the published CPU clock? When chips are manufactured by a specific process, the manufacturer tests samples for stability and performance. Such performance tests generally show a distribution of stability vs. clock speed. The manufacturer then selects a maximum clock speed at which the given chip is 'stable' based on some statistical analysis of the distribution. For example, the manufacturer may select a 99th percentile clock, meaning that 99% of the bulk population is predicted to be stable at the selected clock. Each manufacturer uses its own statistical criteria as well as test methods and definitions of stability. Therefore, comparisons of clock speeds across manufacturers can be misleading.

At present, the two big PC CPU manufacturers are Intel and Advance Micro Devices (AMD). Masses of privately conducted overclocking (see Chapter 2 for a discussion of overclocking) tests have shown that the AMD chips consistently remain stable at speeds much higher than 'published.' This does not itself suggest that AMD chips are 'superior,' but simply suggests that AMD is more conservative in assigning a published clock speed. An example is a 1000 MHz chip reportedly stable at over 2000 MHz; it is not uncommon at all to see reports of 1200 or 1333 MHz AMD CPUs running at 1700 MHz with 'no complaints.' Intel chips have generally been produced to resist overclocking, so the data is less extensive; however, the data do suggest that Intel chips are rated much closer to a practical stability limit.

That AMD has implicitly supported the overclocking users explains much of why AMD seems to be favored among the high performance gaming community. As outlined in Chapter 2, overclocking can be achieved in two ways: increasing the Front Side Bus (FSB) clock and/or increasing CPU multiplier. For several years, Intel chips have had the multiplier 'set and sealed,' so this parameter cannot be adjusted. AMD, on the other hand, has allowed not only CPU multiplier but other technical parameters to be configured by the user. This may contribute to a number CPU's being destroyed by the user, but AMD has allowed experimentation for those willing to risk the life of their CPU.

Before leaving the subject of CPU clock speed, a quick note on AMD's 'newer' CPU model naming scheme is warranted. Until the release of the Athlon XP CPU's, AMD simply gave a model as clock speed, such as the Athlon 1200 (meaning an Athlon core designed to run at 1200 MHz). However, AMD realized the marketing disparity when Intel released, for example, the P4 1700, at the time when AMD chips were typically around 1400 MHz. Therefore, AMD changed the model name scheme. The model name Athlon XP 1700+ does not imply a chip formally running at 1700 MHz, but communicates an *effective* speed comparison to *Athlon-Thunderbird* chips. In other words, the Athlon XP 1700+ had 1470 MHz clock, but benchmarks at or above where a 1700 MHz Thunderbird would perform. There are two key points here. First, AMD Athlon XP tends to benchmark at or above *higher clock* P4's This suggests AMD chips have more efficient instructions per cycle. Specifically, there are examples of AMD XP 1700+ (with 1470 MHz clock) outperforming P4 1900 (and other similar results through the P4 3000's). Second, the 1700+ designation is a performance comparison to *AMD* chips, not Intel.

Intel and AMD show other differences as well. While both support the same *basic* instruction set, their advanced instruction sets are different. This may not matter to code compiled with a general compiler that only uses the basic set. If a compiler is known to compile optimized code for one particular chip (Intel P4, for example), chances are that these advanced instructions generated. The code will run on other chips, but emulation routines may be employed which are slower. However, even if advanced instructions sets are *not* used, the most efficient scheduling of op-codes will be different for the different chips. For maximum performance, it is important to use a compiler that will optimize for the processor used.

Within the basic instruction set, the different chips favor different tasks. Older AMD processors were typically viewed to outperform their Intel counterparts at floating point math. On the other hand, Intel often got the nod in terms of memory transfers. When AMD introduced 266 MHz Double Data Rate (DDR) memory several years ago, Intel's advantage was briefly lost (the Pentium III supported a 100 MHz bus clock). Intel's initial answer was the 400 MHz QDR (Quad Data Rate) RAMBUS memory associated with the Pentium 4. Tests comparing the newest generation AMD to Intel CPU's are complicated at best due to different cache sizes and memory bandwidth issues. The reader is encouraged to research such tests by searching "AMD vs Intel," "AMD Benchmark" or similar keywords in a major search engine.

A final comment should be made regarding heat. High performance CPU's running in high performance systems generate a significantly larger amount of heat than home-user, consumer class systems. This also adds to the cost of a high performance system; robust cooling capacity is required. AMD CPU's in particular are known to 'run hot,' which is a specific trade-off to the high performance floating point processor. Further, cooling equipment supplied by a manufacturer may be inadequate for high performance systems, and many high performance system users revert to after-market components. Such after market components include heat-sinking fans (HSF) with high flow fans, case fans (secondary to primary cooling supplied on the CPU itself), electro thermal, water and glycol based liquid cooling. In a related idea, high performance computers are often run without covers to allow improved passive airflow. Indeed, several of the author's systems run with covers removed with an ordinary home box fan providing secondary airflow.

B.4 MAIN BOARD

The best CPU money can buy will be rendered 'low performance junk' by a low performance main board (also called a motherboard or the slang mobo'). While it may be true that the CPU is the 'brain' of the system, the main board is the nervous system; without high performance connectivity with the rest of the system, the best components cannot perform at peak capacity. It is with the main board that the author believes most computer buyers hamper system performance. PC main boards cover a broad range of prices, from $20 or $30 to $250 or more. Buyers looking only at price often get a low-end main board, and then spend extra money on a 'high performance' CPU that will never perform at its potential.

Hardware sellers place main boards into one of two broad categories: integrated and non-integrated. Integrated boards have peripheral devices, such as parallel ports, serial ports, modems, sound adaptors, video adaptors, network interfaces, etc., 'built into' the board. These boards are generally low-end boards; none of the devices present, including the main board itself, is of very high quality. Integrated boards tend to appeal to home users who favor apparent system simplicity over real performance. For example, consider the marketing copy: "many system components, such as modem, built right-in, saving you money." Integrated boards including common peripherals can be purchased for less than $60; this saves more money since the additional components do not need to be purchased. This is how mass producers can sell systems for less than $500, or even give away 'free' systems along with Internet sign-up commitments. It is typically possible to 'disable' the integrated peripherals and add cards to board slots thus 'upgrading' the board to higher performance components; the underlying system is still a low-end board, however.

Non-integrated boards do not have common peripherals 'built-in;' exceptions are the IDE controller for IDE disk drives and possibly basic I/O ports (parallel, mouse, keyboard, serial and USB). Non-integrated boards allow the coupling of a quality board with other quality components. All high performance PC main boards are non-integrated. Non-integrated boards themselves cost more, and also require the additional purchase of other system components, such as modems, network interfaces, sound adaptors and video adaptors. Many manufacturers produce both integrated and non-integrated boards to service both the low performance and high performance markets.

The defining characteristic of a main board is the chipset employed, which is in turn specified by the CPU for which the board is designed. In other words, boards designed for AMD chips will not work with Intel chips, and vice versa. This is why it is important to select the CPU *first*. If one does not have a particular CPU preference, one can choose the main board first and simply use the CPU that is supported. The functional structure of the chipset is outlined in Chapter 2.

It should be emphasized that many main boards that use the *same* chipset may have significant performance differences. Indeed, the side-by-side quantum chemistry computation comparison mentioned above using the same physical CPU, hard drive and memory was with two boards both having the same chipset; this test resulted in the best case board running over twice as fast as the worst case board. The physical design and construction of the board has definite performance consequences. Modern computer equipment involves complicated high frequency signaling, and paths taken by these signals can be good or bad. Just because a board uses a particular chipset and supports a given CPU does not mean the board is 'high performance.'

Another characteristic specification of the main board is the supported Front Side Bus (FSB) speed (which is generally determined by the chipset). This clock parameter will have a large impact on data transfers *outside* the CPU (ie, it will not matter to CPU register to register transfers for CPU's running at a given frequency). It will do little good to have a CPU capable of interfacing to a 266 MHz bus if the main board only supports 100 MHz. One should also remember that the documented FSB clock of a given board is to some extent similar to the published clock of a given CPU. The board itself may 'allow' faster FSB clocks, but the manufacturer does not 'support' this capability. An example is an ABIT KT7E board documented to run 100 MHz FSB; the author has a KT7E that has been stable at 133 MHz in agreement with many similar reports on the Internet; some KT7E's have been driven to 147 MHz (the maximum for this board). With the factory CPU multiplier of 12, a 133 MHz FSB clock would have the CPU running at 1596 MHz – essentially a "1200 MHz" chip running at 1600 MHz.

The BIOS included on a board also varies extensively between manufacturers. The BIOS must support certain basic functional tasks in a somewhat standardized way, but actual implementation varies. Further, higher end boards generally allow BIOS upgrades and more user configuration options in the bus timing and other hardware specific parameters.

Clock speed is not the only parameter that determines CPU and system performance. There is a host of technical specifications that ultimately determine the speed and stability of a given system, which include CPU core voltages, memory wait states, CAS latency, write-backs, etc. By allowing many of these system board parameters to be configured by the user, a user can 'tune' the system for the particular CPU and other hardware present, rather than taking statistically predicted parameters. Of course, this tuning must be done carefully since it makes little sense to tune a system to parameters that destroys the hardware. Further, any adjustments to system level parameters should be extensively tested for stability. Specifically, if numerical routines are used, floating point accuracy should be verified if the system is configured differently from the manufacturer's defaults. However, even without such technical tuning, boards that allow these parameters to be configured are generally faster than boards that do not.

Quality main boards have various features. Some of these may be more aesthetic than functional, but they exist nonetheless. For example, boards have different layouts. To a casual user, this may make no difference; layout will certainly influence the integrity of the complex high frequency signals that must propagate around the board. Further, layout features such as location of on-board capacitors being in close proximity to the CPU socket can inhibit the addition of an aftermarket HSF. Systems may place memory slots where they may be inaccessible due to the arrangement of drive bays, or worse, CPU locations that inhibit airflow. Some boards provide additional features such as additional fans for the chipset.

Other features include the number and location of fan and other headers (a header is a 'plug' on the board to power and/or control a device, usually fans, LED's and switches). Different boards offer different numbers (and kinds) of expansion slots. Of course, the term 'expansion slot' may be misleading when considering a non-integrated board. Some boards support larger memory spaces than others, and higher or lower Advanced Graphics Port (AGP) bandwidths. Many board features can be taken as they come, but others (such as the AGP or FSB clock) will determine the nature of other hardware (such as video adaptor or memory).

Some current manufacturers of high performance boards include Abit, Asus, Iwill,

Micron and Tyan. At the time of this writing, the high performance models offered from these manufacture's range from about $100-$250, and include various combinations of the features mentioned above.

B.5 MEMORY

The quantity of system RAM effects performance. In general, it is assumed 'more is better,' but this is often not the case with older 'home user' grade operating systems (Windows ME, for example). Certainly if one is applying the Look-Up Table techniques discussed in Chapter 9, a large memory space is required for high performance applications. Like CPU clock selection, the quantity of RAM is more often a matter of affordability. In general, most sellers of RAM scale the price so the cost per MB decreases when larger chips are purchased.

RAM is designed to operate at specific FSB speeds. For example, 100 MHz RAM will not function properly on a board running at 133 MHz. Further, RAM has several specifications stating the address selection latencies. Obviously, 10 ns RAM is faster than 50 ns RAM, meaning data is reliably read in 1/5 the time. One can often check timing of memory read/writes in the main board Set-Up to tune for the RAM that is installed.

System memory can exert numerous influences on the overall performance. From the hardware perspective, there are currently three major memory subsystem designs: standard SDRAM, DDR SDRAM and RAMBUS. Manufacturers of high quality RAM include Samsung, Micron, Corsair and XTreme.

B.6 DISK DRIVES

Disk I/O will often be a significant bottleneck in performance applications, especially those requiring extremely large datasets. Including high quality drives in a system enhances performance noticeably. In addition, the higher quality drives are generally more reliable, which is really the *first* criterion to apply to these critical storage media.

Disk drives are characterized by their interface, speed, seek time, and bandwidth. As with CPU and memory, drives are typically selected based on budget – the best drive available for the budgeted amount is purchased. Secondary to these specifications are capacity; it is assumed that one considers drive capacities that are readily available in a variety of specifications.

The two common drive interfaces include Integrated Drive Electronics (IDE) and Small Computer Systems Interface (SCSI). IDE is common in less expensive systems and traditionally was slower than SCSI; bandwidth is specified as ATA 33, ATA 66, ATA 100 and ATA 133. However, ATA 100 IDE drives operate at near SCSI performance. SCSI drive bandwidths are termed FAST, WIDE and ULTRAWIDE. SCSI devices are also more reliable in systems requiring high availability: they intrinsically support mirroring, striping and hot swapping. IDE drives can be used in mirroring and striping modes provided Redundant Array of Independent Disks (RAID) support is included in the IDE interface. Such RAID support must exist on the main board, and common IDE RAID controllers support RAID 0 (striping), RAID 1 (mirroring) and RAID 0+1 (striping + mirroring).

Many contemporary computers have CD-Rom drives (or writers) or even DVD drives. There is one cautionary note with regard to these IDE devices. They should *never* exist on the same IDE channel as a hard drive needed for high performance (in a master/slave configuration).

SCSI drives are generally higher performance and more reliable. In addition, SCSI channels can hold more drives than IDE ones, which are generally limited to two drives per channel. However, SCSI components are typically more expensive than IDE drives. High performance drive manufactures include IBM, Western Digital and Maxtor.

B.7 CASES AND POWER SUPPLIES

Though many casual users may not realize it, the case containing the computer is a significant component to be considered. Here only ATX style cases are considered, but rack mount cases are also important in performance computing. The case typically includes the power supply, drive (and other peripheral) bays, and bays for fans. A high performance computer running CPU intensive software will draw more power than a home-user computer being used to surf the Internet. Ripple currents (and other power instabilities) not only lowers CPU performance, but also can lead to hardware failures. It is recommended that an oversize power supply be used on any computer that will be heavily used. This is needed to drive heavily loaded CPU, disk drives, multiple fans, high performance memory, etc.

Modern 'performance' cases include multiple case fans, or at least places to mount them. With high performance CPU's and other components running 'full time,' excess heat must be removed from the computer. This is easily overlooked when buying department store computers for performance use.

B.8 VIDEO ADAPTOR

Since this book focuses on floating point intensive computation, a detailed review of video hardware is not given here; indeed, an entire book can be written outlining differences and merits of current high-end video hardware. Having mentioned this, if a system is being purchased a pure computational system, an inexpensive video adaptor can be added. It should be noted that high performance video hardware does generate significant heat; eliminating this component if not needed will help the overall heat problem.

B.9 OTHER DEVICES

Other devices include everything not mentioned previously in this Appendix. This discussion cannot possibly address all possible hardware one may connect to the computer. Care should be exercised when attaching peripherals to a computer built with high performance in mind. One might emphasize that a purpose built computational system is better *not* utilized for data acquisition and other 'ancillary' tasks.

B.10 A FINAL COMMENT ON COST

While it might be thought from the preceding discussion that the author is advocating only *expensive* computers, I made the argument near the beginning of this Appendix that performance systems are not *that much* more expensive. To illustrate this point, consider that at the time of this writing, mass-produced, department store home-user grade systems typically sell for around \$600-\$1000, depending on features, bundled software, etc.

The author recently custom built a performance, purely computational computer for \$960. This system used the ABIT KX7-333R board, AMD Athlon 2100+ CPU, 512 MB of 6ns, PC2700 XTreme DDR RAM, two 60 GB ATA 133 Maxtor hard drives (in RAID 1 configuration) and a 3Com Network Card. No monitor or keyboard was purchased (access to the system is via LAN), and Open Source GNU licensed software (including the OS) was used. This system benchmarks well above the home-user grade systems of similar price. In fact, it benchmarks favorably against a mainframe architecture: an early *ab initio* quantum chemistry benchmark showed an execution time of 3.7 seconds, compared to 115 seconds on an RS/6000-350 machine (that is 31 times faster, where the processor clock difference accounts only for a factor of 6). The big savings were, of course, in monitor and OS (Open Source vs Windows), but even had these been purchased, the price would have been approximately \$1500. The reader is encouraged to remember that high performance is not *much* more expensive than nearly useless (in a performance sense) junk, and a system should be evaluated based on *value* not on price tag.

Appendix C:

Contents of the Included CD-Rom

In addition to the demo programs used to test the algorithms in this book, the CD-Rom contains a number of How-To's and tools. This appendix summarizes the contents.

Top Level Directories

Applications These are 'complete' working applications written by the author. They are small, simple educational demos.

Compilers Demo versions of compilers provided by developers

Demos Demo programs used in this book. The source code and project files are included (as well as executable images for the Windows versions). Each chapter's demos are in a separate directory.

Math Libraries Actual libraries or information about libraries for new developers to quickly get started with basic code.

Tutorials Third party help files and How-To's.

Appendix E This Appendix contains the source code listings for all the book's demos in BASIC. This is provided for programmers not wishing to follow the source code in C in the book. Appendix E can be printed so the reader can follow the source code without having to constantly flip to the Appendix at the end of the text.

Applications

There are two demo applications included, and both are for the Windows platform. The author plans to port these applications to Linux, in which case they will be available for free download from the author's web site (www.dsbscience.com).

MCGas This demo program is the small Monte Carlo simulation application developed for use as a General or Physical Chemistry 'lab' exercise. The program is written in Visual Basic and includes 'real' Lennard Jones energy functions for helium, argon, xenon, hydrogen, nitrogen, oxygen, methane and carbon tetrachloride. The Readme.txt file contains more information, and the program includes a Windows html 'Help' file. The complete source code for MCGas Demo is included in Appendix D, since this program incorporates numerous optimizations discussed in this book.

VolSbmSph This demo program models a floating sphere and was written for use in a basic Physics laboratory course. The user can change the mass and radius of the sphere as well as the density of the fluid, and observe the effect on the depth of floatation.

Compilers

VectorC_PC_Demo_Setup_2.0.1.zip

This is the demo version of the VectorC high performance compiler by CodePlay. This compiler produces SIMD code natively and includes numerous other optimizations for producing fast code.

Miracle C Compiler Info regarding the shareware C compiler for Windows. Actually, this compiler is very good for learning C, since it provides little feedback for errors and warnings. Miracle C forces the programmer to study the code when something does not work properly.

Demos

BASIC This directory contains the Windows and Generic versions of the demo programs used in this book. The Windows versions were written for Visual Basic and include all project files needed to compile the demos. Executables are included for the Windows versions.

C This directory contains the C/C++ versions of the Demos for Windows and Linux.

Linux The Linux Version of the Demos. In the Linux directory, there is an INSTALL file that contains instructions for using the makeall.sh script that will compile all the demos from the command line.

Windows There are two Windows versions: MFC C++ and Console (mostly straight C). The MFC applications were created with the Visual C++ App Wizard; the working functions are mostly straight C, with some exceptions. Project files and executables are included. For the straight C console applications, the executables and C source files are included.

Math Libraries

GenMath This is a very small library of general math routines written by the author. This library is provided simply for new programmers or those without larger, more developed libraries to get started with numerical programming. The library is written completely in straight C and should be platform portable (hence the designator 'general'). The library will be updated periodically; updates will be provided free on the author's web site (www.dsbscience.com).

GenMathDoc The documentation for the GenMath library.

libSIMD This is info regarding the libSIMD SourceForge project headed by Iain Nicholson. The goal of this project is to produce an SIMD optimized library for common math routines of interest to scientists.

SciMath This is info regarding the SciMath library, developed by ASA, Inc (included by permission).

Tutorials

ASP Page from the ASP Emporium about using the GetObject() Function, including a link to a page about using the WinMgmts moniker. The WinMgmts moniker is needed to use the WMI functions in Windows, useful for remote process execution as used for developing low complexity parallel applications for office networks.

Assembly Language The "Using Inline Assembly with gcc" How-To written by Clark L. Coleman, Brendan Underwood and colin@nyx.net.

General Linux How-To's

3D Modelling How-To by Dave Jarvis.
Benchmarking How-To by Andr D. Balsa
gcc How-To by Daniel Barlow
Program Library How-To by David A. Wheeler
README Information regarding from where these How-To's were downloaded.
Scientific Computing with GNU Linux How-To by Manoj Warrier

Parallel Processing

Beowulf How-To by Jacek Radcjewski and Douglas Eadline
Cluster How-To by Ram Samudrala
Open Mosix How-To by Kris Buytaert
Parallel Processing How-To by Hank Dietz

PHP

php_manual_en.chm
Windows html 'Help' file manual for php by Stig Sæther Bakken and Egon Schmid
PHP How-To by Alavoor Vasudevan

Appendix D: Complete Code Listing of MCGas Demo Version

Note: This listing does not include the code for implementation of the MCGas.chm HTML Help. This listing is for the frmMCGasMain.frm source file.

```
Begin VB.Form frmMCGasMain
    Caption             =    "MCGas Monte Carlo Gas Simulation"
    ClientHeight        =    6465
    ClientLeft          =    165
    ClientTop           =    450
    ClientWidth         =    5625
    LinkTopic           =    "Form1"
    ScaleHeight         =    6465
    ScaleWidth          =    5625
    StartUpPosition =    2   'CenterScreen
    Begin VB.CommandButton Command1
        Caption             =    "Simple Ising Magnet"
        Height              =    495
        Left                =    3360
        TabIndex            =    19
        ToolTipText         =    "Simulate Simple Ising Magnet
using the entered Temperature"
        Top                 =    3600
        Width               =    1455
    End
    Begin VB.PictureBox Picture1
        Height              =    1335
        Left                =    0
        ScaleHeight         =    400
        ScaleMode           =    0   'User
        ScaleWidth          =    400
        TabIndex            =    18
        Top                 =    3360
```

```
      Visible              =    0     'False
      Width                =    1335
   End
   Begin VB.CommandButton cmdCompute
      Caption              =    "Begin Full Calculation"
      Height               =    495
      Left                 =    1440
      TabIndex             =    16
      ToolTipText          =    "Simulate Gas using the
entered parameters"
      Top                  =    3600
      Width                =    1575
   End
   Begin VB.Frame Frame2
      Caption              =    "Energy Function"
      Height               =    1095
      Left                 =    3000
      TabIndex             =    13
      Top                  =    2040
      Width                =    1935
      Begin VB.OptionButton Option4
         Caption              =    ""
         Enabled              =    0     'False
         Height               =    375
         Left                 =    120
         TabIndex             =    15
         Top                  =    600
         Visible              =    0     'False
         Width                =    1575
      End
      Begin VB.OptionButton Option3
         Caption              =    "Lennard-Jones"
         Height               =    255
         Left                 =    120
         TabIndex             =    14
         Top                  =    240
         Value                =    -1    'True
         Width                =    1695
      End
   End
   Begin VB.Frame Frame1
      Caption              =    "Geometry"
      Height               =    1095
      Left                 =    600
      TabIndex             =    10
      Top                  =    2040
```

```
        Width                 =    1455
        Begin VB.OptionButton Option2
            Caption                =    "3 D"
            Enabled                =    0      'False
            Height                 =    375
            Left                   =    480
            TabIndex               =    12
            Top                    =    600
            Width                  =    855
        End
        Begin VB.OptionButton Option1
            Caption                =    "2 D"
            Height                 =    255
            Left                   =    480
            TabIndex               =    11
            Top                    =    240
            Value                  =    -1     'True
            Width                  =    735
        End
    End
    Begin VB.TextBox Text4
        Alignment             =    1   'Right Justify
        Height                =    285
        Left                  =    4680
        TabIndex              =    9
        Text                  =    "30000"
        Top                   =    1080
        Width                 =    735
    End
    Begin VB.TextBox Text2
        Alignment             =    1   'Right Justify
        Height                =    285
        Left                  =    1560
        TabIndex              =    6
        Text                  =    "40"
        Top                   =    1440
        Width                 =    735
    End
    Begin VB.TextBox Text1
        Alignment             =    1   'Right Justify
        Height                =    285
        Left                  =    1560
        TabIndex              =    3
        Text                  =    "100"
        Top                   =    840
        Width                 =    735
```

```
End
Begin VB.ComboBox Combo1
    Height              =   315
    ItemData            =   "frmMCGasMain.frx":0000
    Left                =   1200
    List                =   "frmMCGasMain.frx":0002
    Style               =   2  'Dropdown List
    TabIndex            =   1
    Top                 =   240
    Width               =   1815
End
Begin VB.Label Label8
    Height              =   1095
    Left                =   1560
    TabIndex            =   17
    Top                 =   4440
    Width               =   2295
End
Begin VB.Label Label7
    Caption             =   "Number of Configurations:"
    Height              =   495
    Left                =   3360
    TabIndex            =   8
    Top                 =   960
    Width               =   1095
End
Begin VB.Label Label5
    Caption             =   "Angstroms"
    Height              =   255
    Left                =   2520
    TabIndex            =   7
    Top                 =   1440
    Width               =   735
End
Begin VB.Label Label4
    Caption             =   "Box Size:"
    Height              =   375
    Left                =   240
    TabIndex            =   5
    Top                 =   1440
    Width               =   1095
End
Begin VB.Label Label3
    Caption             =   "Kelvin"
    Height              =   255
    Left                =   2520
```

```
      TabIndex          =     4
      Top               =     840
      Width             =     495
End
Begin VB.Label Label2
      Caption           =     "Temperature:"
      Height            =     375
      Left              =     240
      TabIndex          =     2
      Top               =     840
      Width             =     1095
End
Begin VB.Label Label1
      Caption           =     "System:"
      Height            =     375
      Left              =     240
      TabIndex          =     0
      Top               =     240
      Width             =     735
End
Begin VB.Menu mnuFile
      Caption           =     "File"
      Index             =     1
      Begin VB.Menu mnuOpenEvC
            Caption           =     "Open Energy vs. Config"
            Enabled           =     0     'False
            Index             =     3
      End
      Begin VB.Menu mnuSaveEvC
            Caption           =     "SAVE Energy vs. Config"
            Enabled           =     0     'False
            Index             =     2
      End
      Begin VB.Menu mnuPrintEvC
            Caption           =     "Print Energy vs. Config"
            Enabled           =     0     'False
            Index             =     4
      End
      Begin VB.Menu mnuB1
            Caption           =     "-"
            Index             =     5
      End
      Begin VB.Menu mnuOpenPC
            Caption           =     "Open Particle Config"
            Enabled           =     0     'False
            Index             =     6
```

```
    End
    Begin VB.Menu mnSavePC
        Caption             =    "SAVE Particle Config"
        Enabled             =    0     'False
        Index               =    7
    End
    Begin VB.Menu mnuPrintPC
        Caption             =    "Print Particle Config"
        Enabled             =    0     'False
        Index               =    8
    End
    Begin VB.Menu mnuB2
        Caption             =    "-"
        Index               =    9
    End
    Begin VB.Menu mnuOpenRDF
        Caption             =    "Open RDF"
        Enabled             =    0     'False
        Index               =    10
    End
    Begin VB.Menu mnuSaveRDF
        Caption             =    "Save RDF"
        Enabled             =    0     'False
        Index               =    11
    End
    Begin VB.Menu mnuPrintRDF
        Caption             =    "Print RDF"
        Enabled             =    0     'False
        Index               =    12
    End
    Begin VB.Menu mnuB3
        Caption             =    "-"
        Index               =    13
    End
    Begin VB.Menu mnuExit
        Caption             =    "Exit"
        Index               =    13
    End
End
Begin VB.Menu mnuOptions
    Caption             =    "Options"
    Index               =    14
    Begin VB.Menu mnuPlotEvC
        Caption             =    "Plot Energy vs. Config"
        Enabled             =    0     'False
        Index               =    15
```

```
        End
        Begin VB.Menu mnuPlotPC
            Caption             =    "Plot Particle Config"
            Enabled             =    0    'False
            Index               =    16
        End
        Begin VB.Menu mnuPlotPDF
            Caption             =    "Plot RDF"
            Enabled             =    0    'False
            Index               =    17
        End
    End
    Begin VB.Menu mnuTools
        Caption             =    "Tools"
        Index               =    22
        Begin VB.Menu mnuToolMem
            Caption             =    "Memory Requirement
Estimator"
            Index               =    21
        End
    End
    Begin VB.Menu mnuHelp
        Caption             =    "Help"
        Index               =    18
        Begin VB.Menu mnuMCHelp
            Caption             =    "MC Gas Help"
            Enabled             =    0    'False
            Index               =    19
        End
        Begin VB.Menu mnuB4
            Caption             =    "-"
            Index               =    20
        End
        Begin VB.Menu mnuAbout
            Caption             =    "About MC Gas"
            Index               =    21
        End
    End
End
Attribute VB_Name = "frmMCGasMain"
Attribute VB_GlobalNameSpace = False
Attribute VB_Creatable = False
Attribute VB_PredeclaredId = True
Attribute VB_Exposed = False
'-------- MCGas ----------'
            Monte Carlo Gas Simulator
```

```
'
' This program originally written by:
'
'    John S. Riley
'    DSB Scientific Consulting
'    dsbscience@cetlink.net
'
' This program performs basic Metropolis
' Monte Carlo in two different ways.  The
' first is a simple Ising Magnet with
' nearest neighbor interactions modeled.  '
' The second uses the Lennard Jones pair
' energy for several 'real' systems.
'
'___________________'


Option Explicit
'global declarations
'constants
Const cEStep As Long = 10                'multiplier for
                                         'distance mesh
Const kb As Double = 1.381E-23           'Boltzmann's constant
Const Sqrt2 As Double = 1.414213562 'square root of two
Const cPi As Double = 3.141592654    'pi

'longs
Dim NumConfigs          As Long 'number of configuration
Dim NumPart             As Long 'number of particles
Dim BoxSide             As Long 'size of box
Dim Xk_1                As Long 'Xk, Xk_1, a and m are used
                                'to compute random numbers
Dim Xk                  As Long
Dim A                   As Long
Dim m                   As Long
Dim X()                 As Long 'particle x coordinates
Dim Y()                 As Long 'particle y coordinates
Dim r()                 As Long 'distance Look Up Table

'singles
Dim m_1                 As Single '1/m, to avoid division

'doubles
Dim Temp                As Double 'temperature
Dim beta                As Double '1/temperature
Dim E()                 As Double 'energy Look Up Table
```

```
'strings
Dim Msg1                  As String 'some display messages
Dim Title1               As String

'entry function; performs basic initializations
Private Sub Form_Load()

    'add systems to list box
    With Combo1

        .AddItem "He", 0
        .AddItem "Ar", 1
        .AddItem "Xe", 2
        .AddItem "H2", 3
        .AddItem "N2", 4
        .AddItem "O2", 5
        .AddItem "CH4", 6
        .AddItem "CCl4", 7

        .Text = .List(0)

    End With

    'preload the graphical display form
    Load frmPlotGas

    'initialize the display form
    frmPlotGas.Height = 5000
    frmPlotGas.Width = 5000

    'assign some message strings
    Msg1 = "The entered value for the number of
configurations "

    Msg1 = Msg1 & "exceeds 500,000, so the number of
configurations "

    Msg1 = Msg1 & "has been set to 500,000."

    Title1 = "Number of Configurations Too Large"

    'initialize random function parameters
    A = 25
    m = 2 ^ 24
    m_1 = 1 / m
```

```
     'seed the inline random number function
     Xk = 20000

End Sub

' performs Metropolis Monte Carlo on real systems
' using Lennard Jones energy function.
' The function uses Look Up Tables for both the
' energies and the distances.
Private Sub cmdCompute_Click()

     'local declarations
     'longs
     Dim SystemIndex      As Long 'list box index to
                           ' physical system
     Dim dummy             As Long 'return value for
                           ' MsgBox calls
     Dim MaxEIndex        As Long 'maximum energy index
     Dim EIndex            As Long 'energy array counter
     Dim RIndexX           As Long 'distance array x
                           'counter
     Dim RIndexY           As Long 'distance arran
                           'y counter
     Dim InitialX          As Long 'x for initial
                           'configuration
     Dim InitialY          As Long 'y for initial
                           'configuration
     Dim PartCounter      As Long 'particle counter
     Dim ItCounter         As Long 'configuration counter
     Dim i                   As Long 'particle i index
     Dim J                   As Long 'particle j index
     Dim DeltaX             As Long 'delta-x between
                           'two particles
     Dim DeltaY             As Long 'delta-y between
                           'two particles
     Dim OldX                As Long 'x of selected particle
                           'before moved
     Dim OldY                As Long 'y of selected particle
                           'before moved
     Dim NewX                As Long 'new x of selected
                           'particle
     Dim NewY                As Long 'new y of selected
                           'particle
     Dim SelectedPart     As Long 'selected particle
                           'index
     Dim RejectedPosition As Long 'number of rejected
```

```
                              'configurations
    Dim R0                   As Long 'interparicle distance
                             'of maximum attraction
    Dim iSigma               As Long 'LJ sigma scaled by
                             'distance mesh
    Dim MaxDeltaX            As Long '2x maximum allowable
                             'step in x direction
    Dim MaxDeltaY            As Long '2x maximum allowable
                             'step in y direction

    'singles
    Dim RandomVal            As Single 'computed pseudo
                                'random number

    'doubles
    Dim Depth                As Double 'LJ well depth
    Dim Sigma                As Double 'LJ sigma parameter
    Dim Sigma_6              As Double 'LJ sigma^6
    Dim Sigma_12             As Double 'LJ sigma^12
    Dim Depthx4              As Double 'LJ depth * 4
    Dim MaxR                 As Double 'maximum distance
                                'within box
    Dim RVal                 As Double 'real distance for
                                'computation of energy
    Dim R6                   As Double 'distance^6 to assign
                                'E Look Up Table
    Dim R12                  As Double 'distance^12 to assign
                                'E Look Up Table
    Dim Rx                   As Double 'real delta-x to
assign
                           'R Look Up Table
    Dim Ry                   As Double 'real delta-y to
assign
                                'R Look Up Table
    Dim TotalEnergy          As Double 'total system energy;
                                'sum of pair energies
    Dim deltaEnergy          As Double 'energy change after
                                'single particle moved
    Dim OldDeltaEnergy       As Double 'single particle-rest
                                'of system energy
                                'before move
    Dim NewDeltaEnergy       As Double 'single particle-rest
                                'of system energy
                                'after move
    Dim CompareVal           As Double 'value to test move
```

```
'initialize the display form
frmPlotGas.Height = 5000
frmPlotGas.Width = 5000

'display message
Label8.Caption = "Initializing ..."

'let windows catch up
DoEvents

'get user inputs
NumConfigs = CLng(Val(Text4.Text))

'prevent overreaching...
If NumConfigs > 500000 Then

      NumConfigs = 500000
      dummy = MsgBox(Msg1, vbOKOnly, Title1)

End If

'get boxside from text box
BoxSide = CLng(Val(Text2.Text))

'get temperature from text box
Temp = Val(Text1.Text)

'obey 3rd law, prevent division by 0 kelvin
If Temp < 0.1 Then Temp = 0.1

'compute beta; Depth contains kb
beta = 1 / Temp

'get system from list box
SystemIndex = Combo1.ListIndex

'assign particle system
Select Case SystemIndex

Case 0        'helium

      Depth = 10.22
      Sigma = 2.58

Case 1        'argon
```

```
        Depth  =  124
        Sigma  =  3.42

    Case 2          'xenon

        Depth  =  221
        Sigma  =  4.1

    Case 3          'diatomic hydrogen

        Depth  =  37
        Sigma  =  2.23

    Case 4          'diatomic nitrogen

        Depth  =  91.5
        Sigma  =  3.68

    Case 5          'diatomic oxygen

        Depth  =  118
        Sigma  =  3.46

    Case 6          'methane

        Depth  =  148.2
        Sigma  =  3.82

    Case 7          'carbon tetrachloride

        Depth  =  327
        Sigma  =  5.88

    End Select

    'calculate some needed values:
    'need sigma as a Long, scaled to distance mesh
    iSigma = CLng(Sigma * 10)

    'index the distance of maximum interaction
    R0 = CLng(Sigma * 11.224)

    'depth parameters assume LJ eq'n has 4*Depth
    Depthx4 = Depth * 4

    'compute sigma^6
```

```
Sigma_6 = Sigma ^ 6

'compute sigma^12
Sigma_12 = (Sigma_6) ^ 2

'convert boxside to distance mesh
BoxSide = BoxSide * cEStep

'compute the number of particles that
'fill half the box
NumPart = (BoxSide \ R0 + 1) * (BoxSide \ (2 * _
          R0) + 1)

'set 2x maximum particle step
MaxDeltaX = BoxSide \ 4
MaxDeltaY = BoxSide \ 4

'configure graphical display box
frmPlotGas.ScaleTop = BoxSide
frmPlotGas.ScaleHeight = -BoxSide
frmPlotGas.ScaleWidth = BoxSide

'compute the maximum distance inside the box
MaxR = BoxSide * Sqrt2

'compute maximum energy lookup table index
MaxEIndex = (CLng(MaxR) + 1)

'initialize the energy table
ReDim E(MaxEIndex)

For EIndex = 1 To MaxEIndex

     'compute real distance from lut index
     RVal = CDbl(EIndex) * 0.1

     'compute r^6
     R6 = RVal * RVal * RVal * RVal * RVal * RVal

     'compute r^12
     R12 = R6 * R6

     'compute E/4C
     E(EIndex) = (Sigma_12 / R12) - (Sigma_6 / R6)

     'compute E
```

```
        E(EIndex) = E(EIndex) * Depthx4

    Next EIndex

    'compute the maximum index for the distance LUT
    ReDim r(CLng(MaxR) + 1, CLng(MaxR) + 1)

    'initialize the r table
    For RIndexX = 0 To CLng(MaxR)

        'compute real delta-x from x index
        Rx = CDbl(RIndexX * RIndexX)

        For RIndexY = 0 To CLng(MaxR)

            'compute real delta-y from y index
            Ry = CDbl(RIndexY * RIndexY)

            'compute distance for delta-x, delta-y
            r(RIndexX, RIndexY) = CLng(Sqr(Rx + Ry))

        Next RIndexY

    Next RIndexX

    'set initial configuration, 'condensed' at bottom
    'of box
    'first particle goes at 0,0
    InitialX = 0
    InitialY = 0

    'initialize x and y coordinate arrays for particles
    ReDim X(NumPart)
    ReDim Y(NumPart)

    For PartCounter = 1 To NumPart

        'put first particle at 0,0
        X(PartCounter) = InitialX
        Y(PartCounter) = InitialY

        'increment x by r0
        InitialX = InitialX + R0

        'reset x to 0 when row is full
        'and increment y by r0
```

```
        If InitialX > BoxSide Then

            InitialX = 0
            InitialY = InitialY + R0

        End If

    Next PartCounter

    'open display form
    frmPlotGas.Show

    'display initial configuration
    'first be sure display is cleared
    frmPlotGas.Cls

    'draw each particle
    For PartCounter = 1 To NumPart

        frmPlotGas.Circle (X(PartCounter), _
            Y(PartCounter)), R0 / 2

    Next PartCounter

    'inform user that calc. is set-up
    Call MsgBox("", vbOKOnly, "Enter or OK to Begin")

    'display message in label
    Label8.Caption = "Starting Monte Carlo ..."

    'let Windows catch up
    DoEvents

    '***** begin actual monte carlo procedure *****

    'change mouse pointer to hourglass to show "busy"
    'for both forms
    frmMCGasMain.MousePointer = 11
    frmPlotGas.MousePointer = 11

    'zero the rejected counter
    RejectedPosition = 0

    For ItCounter = 1 To NumConfigs

        'pick a random particle
```

```
Xk_1 = (A * Xk) Mod m
RandomVal = CSng(Xk_1 * m_1)
Xk = Xk_1

SelectedPart = CLng(RandomVal * NumPart)

'store the current coordinates of the
'selected particle
OldX = X(SelectedPart)
OldY = Y(SelectedPart)

'**** Try a Move ****

'get a random new x coordinate
Xk_1 = (A * Xk) Mod m
RandomVal = CSng(Xk_1 * m_1)
Xk = Xk_1

NewX = ((RandomVal * MaxDeltaX) - (MaxDeltaX * _
      0.5)) + X(SelectedPart)

'apply boundary condition
If NewX > BoxSide Then

     NewX = NewX - BoxSide

ElseIf NewX < 1 Then

     NewX = NewX + BoxSide

End If

'get a random new y coordinate
Xk_1 = (A * Xk) Mod m
RandomVal = CSng(Xk_1 * m_1)
Xk = Xk_1

NewY = ((RandomVal * MaxDeltaY) - (MaxDeltaY * _
      0.5)) + Y(SelectedPart)

'apply boundary condition
If NewY > BoxSide Then

     NewY = NewY - BoxSide

ElseIf NewY < 1 Then
```

```
            NewY = NewY + BoxSide

      End If

      'zero the energies
      OldDeltaEnergy = 0
      NewDeltaEnergy = 0

      ' compute deltaEnergy
      For PartCounter = 1 To NumPart

            If PartCounter <> SelectedPart Then

                  'delta energy for old position
                  DeltaX = Abs(OldX - X(PartCounter))
                  DeltaY = Abs(OldY - Y(PartCounter))

                  EIndex = r(DeltaX, DeltaY)

                  OldDeltaEnergy = OldDeltaEnergy + _
                             E(EIndex)

                  'delta energy for new position
                  DeltaX = Abs(NewX - X(PartCounter))
                  DeltaY = Abs(NewY - Y(PartCounter))

                  EIndex = r(DeltaX, DeltaY)

                  NewDeltaEnergy = NewDeltaEnergy + _
                             E(EIndex)

            End If

      Next PartCounter

      deltaEnergy = NewDeltaEnergy - OldDeltaEnergy

      '**** test the new configuration ****
      'apply Metropolis Monte Carlo
    'with Boltzmann Weighting
      If deltaEnergy <= 0 Then

            'force 'accept' if new configuration
            'has lower (or equal) energy than old
            CompareVal = 2
```

```
Else

      'otherwise, apply Boltzman distribution
      CompareVal = Exp(-deltaEnergy * beta)

End If

'get a random number to test against
'Boltzmann distribution
Xk_1 = (A * Xk) Mod m
RandomVal = CSng(Xk_1 * m_1)
Xk = Xk_1

'CompareVal = 2 if new energy < old energy
'RandomVal 0-1, so the move is always accepted
'for case of CompareVal =2.  Otherwise,
'Compare is a Boltzmann factor
If CompareVal >= RandomVal Then

      'accept the move weighted by
      'Boltzmann, or if
      'new energy < old energy
      X(SelectedPart) = NewX
      Y(SelectedPart) = NewY

Else

      'reject move if RandomVal is 'outside'
      'Boltzmann factor
      RejectedPosition = RejectedPosition + 1
      X(SelectedPart) = OldX
      Y(SelectedPart) = OldY

End If

'update display every 1000 iterations
If ItCounter Mod 1000 = 0 Then

      'clear the display area
      frmPlotGas.Cls

      'plot all particles in current position
      For PartCounter = 1 To NumPart

            frmPlotGas.Circle (X(PartCounter), _
```

```
                    Y(PartCounter)), R0 / 2

            Next PartCounter

            'let Windows catch up
            DoEvents

        End If

    Next ItCounter

    'audible alert when done
    Beep

    'compute total energy
    TotalEnergy = 0

    For i = 1 To NumPart - 1

        For J = i + 1 To NumPart

            DeltaX = CLng(Abs(X(i) - X(J)))
            DeltaY = CLng(Abs(Y(i) - Y(J)))

            EIndex = r(DeltaX, DeltaY)

            TotalEnergy = TotalEnergy + E(EIndex)

        Next J

    Next i

    'display total energy
    Label8.Caption = "Energy = " & Str(TotalEnergy) & _
                     " Fraction Rejected = " & _
                     Str(RejectedPosition / _
                     NumConfigs)

    'change mouse pointer back to 'normal'
    frmMCGasMain.MousePointer = 1
    frmPlotGas.MousePointer = 1

End Sub

'This function does the Simple Ising Magnet/
'Lattice Gas Simulation.  The key difference
```

```
'between this simulation and the LJ system is
'that ONLY nearest neighbor interactions are
'considered here; that is, the interaction is
'short range only.
'
'This code is adapted from David Chandler,
'"Introduction to Modern Statistical Mechanics"
'Chapter 6.
Private Sub Command1_Click()

     'local declarations
     'longs
     Dim S              As Long       'selected spin
     Dim i              As Long       'i coordinate
     Dim J              As Long       'j coordinate
     Dim m              As Long       'selected row
     Dim n              As Long       'selected column
     Dim Icount        As Long       'configuration counter
     Dim nplus         As Long       'accepted configurations
     Dim A(22, 22)    As Long       'configuration array

     'singles
     Dim xn             As Single     'accepted fraction
     Dim b              As Single     'argument to compute
                                         'Boltzmann factor

     'doubles
     Dim Temp As Double   'temperature

     'initialize the display area
     frmPlotGas.Height = 5000
     frmPlotGas.Width = 5000

     'get temperature from text box
     Temp = Val(Text1.Text)

     'Obey 3rd Law, avoid 0 kelvin
     If Temp < 0.1 Then Temp = 0.1

     'compute 'beta'; kb is in interaction constant
     Temp = 1 / Temp

     'initial interface pattern
     For i = 1 To 22

          For J = 1 To 10
```

```
            A(i, J) = 1

        Next J

        For J = 11 To 22

            A(i, J) = -1

        Next J

        A(i, 0) = -1
        A(i, 21) = 1

    Next i

    'set display area parameters
    frmPlotGas.ScaleTop = 20
    frmPlotGas.ScaleHeight = -20
    frmPlotGas.ScaleWidth = 20

    'show the display form
    frmPlotGas.Show

    'plot the initial interface
    'clear the plot area
    frmPlotGas.Cls

    'plot the current configuration
    For i = 1 To 20

        For J = 1 To 20

            If A(i, J) = 1 Then frmPlotGas.Circle (i, _
                        J), 0.45

        Next J

    Next i

    'notify that initializations done
    Call MsgBox("", vbOKOnly, "Enter or OK to Begin")

    '****  Begin actual Monte Carlo Procedure ****
    For Icount = 1 To 3000
```

```
'select a random row
m = Int(20 * Rnd + 1)

'select a random column
n = Int(20 * Rnd + 1)

' reverse the selected spin
S = -A(n, m)

'compute argument for Boltzmann factor
b = Temp * S * (A(n - 1, m) + A(n, m - 1) + _
     A(n + 1, m) + A(n, m + 1)) * 2

'test the configuration change
'using Metropolis with Boltzmann Weighting
If Exp(b) >= Rnd Then

     A(n, m) = S
     nplus = nplus + S

     'apply boundary condition
     If n = 1 Then

          A(21, m) = S

     ElseIf n = 20 Then

          A(0, m) = S

     End If

     If m = 1 Then

          A(n, 21) = S

     ElseIf m = 20 Then

          A(n, 0) = S

     End If

End If

'update display every 100th configuration
If Icount Mod 100 = 0 Then
```

```
        'plot the configuration
        'clear the display area
        frmPlotGas.Cls

        'plot the current configuration
        For i = 1 To 20

            For J = 1 To 20

                If A(i, J) = 1 Then frmPlotGas.Circle _
                    (i, J), 0.45

            Next J

        Next i

        'compute the ratio of spins that are reversed
        xn = nplus / 400

        'display results
        Label8.Caption = "n+/n = " & Str(xn)

        'let windows catch up
        DoEvents

    End If

    Next Icount

    'audible alert when finished
    Beep


End Sub

'safer program exit
Private Sub Form_unload(cancel As Integer)

    'local declarations
    Dim countforms As Integer

    'unload all forms except main
    For countforms = Forms.Count - 1 To 0 Step -1

        Unload Forms(countforms)
```

```
    Next

    'unload main form
    Unload Me

    'be sure resources cleared
    Set frmMCGasMain = Nothing

    'terminate now
    End

End Sub

'On click Help_About
Private Sub mnuAbout_Click(Index As Integer)

    'show the form
    frmAbout.Show

End Sub

'On click File_Exit
Private Sub mnuExit_Click(Index As Integer)

    'call the safer exit routine
    Form_unload 0

End Sub

'On click Tools-the memory estimator.  This function
'estimates the memory required for the two LUT arrays
'for energy and distance.  The other memory requirements
'are not included since they will be small compared
'to these two arrays.
Private Sub mnuToolMem_Click(Index As Integer)

    'local declarations
    Dim MaxR                As Long     'maximum r index
    Dim MinMem              As Long     'computed minimum
                                'memory required
    Dim MemE                As Long     'memory needed for
                                'E LUT
    Dim MemR                As Long     'memory needed for
                                'R LUT

    Dim EString             As String   'display string for
```

```
                                    'E memory
    Dim RString             As String   'display string for
                                    'R memory
    Dim MinMemString        As String   'display string for
                                    'required memory
    Dim MsgString           As String   'build string for
                                    'message box

    'compute the maximum distance index
    MaxR = CLng((Val(Text2.Text) * cEStep) * Sqrt2)

    'compute memory required for E array
    MemE = MaxR * 0.008

    'build the E part of the display message
    EString = Format(MemE, "#,###")

    'compute the memory required for the R array
    MemR = MaxR * MaxR * 0.004

    'build the R part of the display message
    RString = Format(Str(MemR), "#,###")

    'compute the total estimated memory required
    MinMem = MemE + MemR

    'build the total part of the display string
    MinMemString = Format(Str(MinMem), "#,###")

    'build the display string and display in message
    'box
    MsgString = "The simulation using the current _
            parameters will require "
    MsgString = MsgString & "(in kilobytes): " & _
            Chr$(13) & Chr$(13)
    MsgString = MsgString & "For Energy Array:     " & _
            EString & Chr$(13)
    MsgString = MsgString & "For Distance Array:  " & _
            RString & Chr$(13)
    MsgString = MsgString & Chr$(13) & Chr$(13)
    MsgString = MsgString & "Total Estimated Minimum _
            kilobytes of Memory Needed: " & _
            MinMemString
    Call MsgBox(MsgString, vbOKOnly, "Lookup Table _
            Minimum Memory Estimator")

End Sub
```

Appendix E: BASIC Listings for the Part I and Part II Demo Programs

Listing 1.1 Pure Scientist Style LJ Energy Computation in BASIC

```
E=0
For i=1 to N

   For j=1 to N

      'avoid division by zero when i=j
      If i<>j Then

         E=E+((σ/sqr((x(i)-x(j))^2+
           (y(i)-y(j))^2)^6 + σ/sqr((x(i)-x(j))^2
           +(y(i)-y(j))^2)^12

      End If

   Next j

Next i
E=E*-C

'avoids double counting the energies since i
'and j are looped over all N
E=E/2
```

Listing 1.2 Short source code for array assignment

```
'SomeValue is defined previously in code
For Count=1 to 10

      R(Count)=SomeValue*Count

Next Count
```

Listing 1.3 Efficient, though longer, source code

```
'SomeValue is defined previously in code

R(1)=SomeValue*1
R(2)=SomeValue*2
R(3)=SomeValue*3
R(4)=SomeValue*4
R(5)=SomeValue*5
R(6)=SomeValue*6
R(7)=SomeValue*7
R(8)=SomeValue*8
R(9)=SomeValue*9
R(10)=SomeValue*10
```

Listing 1.4 Using symmetry to improve Listing 1.1

```
For I=1 to N-1

    For J=I+1 to N

        {code to calculate pairwise energy}

    Next J

Next I
```

Listing 5.3 Floating point arguments with a LONG loop counter

```
'conversion factor for
'degrees to radians,
'divided by 100
Const cDeg2Rad as DOUBLE = π/18000

Dim Counter as LONG
' holder for the computed sin
Dim dummy as DOUBLE

' LONG cannot step 0.01,
' so actual 'degree' * 100
For Counter = 0 to 36000 Step 1

    ' Counter must be converted to
    ' double precision radians
    dummy = sin(CDbl(Counter) * cDeg2Rad)

Next Counter
```

Listing 5.4 Floating-point arguments with floating point loop counter

```
  degrees to radians
Const Deg2Rad as DOUBLE = π/180

Dim x as DOUBLE
Dim dummy as DOUBLE

' x is a Double Precision counter
For x = 0 to 360 Step 0.01

     ' though argument is already a double
     ' it must be converted to radians
     dummy = sin(x * Deg2Rad)

Next x
```

Listing 5.5 Floating-point loop counter without unit conversion

```
Dim Arg1 as DOUBLE
Dim dummy as DOUBLE

' cDeg2Rad defined as in Listing 5.3
For Arg1 = 0 to 2π STEP cDeg2Rad

     dummy = sin(Arg1)

Next Arg1
```

Listing 5.6 LONG loop counter without degree to radian unit conversion

```
' declarations as in Listing 5.3

' counter and step are longs
For Counter = 0 To 62831853 Step 1745

     'sin() gets a double as argument
     dummy = Sin(CDbl(Counter) * cRadStep)

Next Counter
```

Listing 5.7 BASIC function to convert single precision numeric data to a four byte string

```
Private Function sFloat2ByteArray (ByVal qWord As Single)
      As Byte()

      Dim AByte(4) As Byte

      CopyMemory AByte(0), ByVal VarPtr(qWord), 1
      CopyMemory AByte(1), ByVal VarPtr(qWord) + 1, 1
      CopyMemory AByte(2), ByVal VarPtr(qWord) + 2, 1
      CopyMemory AByte(3), ByVal VarPtr(qWord) + 3, 1

      sFloat2ByteArray = AByte

End Function
```

Listing 5.8 Declaration and Initialization of Water Variable using Type MOLECULE

```
Dim Water as MOLECULE

ReDim Water.Particles(3)

' the oxygen nucleus
Water.Particles(1).Charge = 8

' hydrogen nucleus 1
Water.Particles(2).Charge = 1

' hydrogen nucleus 2
Water.Particles(3).Charge = 1
```

Listing 6.1 Example of Code with function call

```
' previous code

A = 1.234
B = X2(A)

'next line of code
'a bunch of code

Private Function X2(Value as Single) as Single
```

```
    Dim Factor as Single
    Factor = 2
    X2 = Value * Factor

End Function
```

Listing 6.2 Scientist "Subroutine" Style of Programming

```
' Function to assign initial
' particle positions
Call Initial_Position

' Function to compute the total
' system energy
Call Calc_Total_Energy

For Configuration=1 to NumberOfConfigurations

      ' Function to move a random
' particle
      Call Move_Particle

      ' Function to compute the
      ' energy change upon moving
      ' the particle
      Call Calculate_Delta_E

      ' Function to test the new
      ' energy against the chosen
      ' distribution function
      Call Test_Move
      ' keep if move okay
      If MoveOK = True Then

            {code to update particle
            positions to keep the new
            configuration}

      End If

' Keep going for selected
' number of configurations
Next Configuration

' Function to output result
Call Print_Energy
```

Listing 6.3 **Coding functionality inline rather than called subroutines**

```
{Actual code to compute Initial Position}
{Actual code to compute Total Energy}

For Configuration = 1 to NumberOfConfigurations

     {Actual code to Move a Particle}

     {Actual code to compute the energy change}

     {Actual code to test the move}

     ' keep if move okay
     If MoveOK = True Then

          {code to update particle
          positions to keep the new
          configuration}

     End If

Next Configuration

{Actual code to Print Output}
```

Listing 6.6 **Calling a function by passing a parameter**

```
Call Calc_Delta_Energy (TheMovedParticle)

Function Calc_Delta_Energy (ParticleMoved as _
          LONG) as Double

     {Code to compute the Lennard-Jones Energy
     change}

End Function
```

Listing 6.7 **Calling a function that uses GLOBAL data**

```
Public ParticleMoved as LONG
Public Delta_E as DOUBLE

{ParticleMoved is set in code}

Public Function Calc_Delta_Energy()
```

```
    {code to compute the Lennard-Jones Energy
    change using the GLOBAL variable
    ParticleMoved, sets the  computed value to
    the GLOBAL variable Delta_E}

End Sub
```

Listing 6.8 Recursive Factorial Function

```
Private Function Recurs_Fact(Arg1 As Long)
        As Double

    Dim answer As Double

    If Arg1 = 1 Then

            Recurs_Fact = 1
            Exit Function

    Else

          answer = Recurs_Fact(Arg1 - 1) * Arg1

    End If

    Recurs_Fact = answer

End Function
```

Listing 6.9 Iterative Factorial Function

```
Private Function Iter_Fact (factval as Long)
          as Double

          Dim fact as Long
          Dim factorial as Double

          fact = 1
          factorial = 1

          Do While fact <= factval
```

```
            factorial = factorial * fact

            fact = fact + 1

        Loop

        Iter_Fact = factorial

Exit Function
```

Listing 7.1 **Simple vector dot product code**

```
c = a_x * b_x  +  a_y * b_y
```

Listing 7.2 **Generalized 2-d vector dot product using arrays**

```
' N is dimensionality; 2 in this case
N = 2

c=0

For Index = 1 to N

    c = c + a (Index) * b (Index)

Next Index
```

Listing 7.3 **2-d vector dot product with the loop unrolled**

```
c = a(1) * b(1) + a(2) * b(2)
```

Listing 7.4 **Generalized rolled-up n x n matrix multiplication**

```
'N is dimensionality
For I = 1 to N

  For J = 1 to N

    c(I, J) = 0

    For k = 1 to N

      c(I, J) = c (I, J) + a(I, k) * b(k, J)
```

```
      Next k
   Next J
Next I
```

Listing 7.5 Explicit rolled-up 4x4 matrix multiplication

```
For I = 1 to 4
   For J = 1 to 4
      c(I, J) = 0
       For k = 1 to 4
          c(I, J) = c(I, J) + a(I, k) * b(k, J)
      Next k
Next J
Next I
```

Listing 7.6 Unrolled 4x4 matrix multiplication

```
'I = 1
C(1, 1) = A(1, 1) * B(1, 1) + A(1, 2) *
          B(2, 1) + A(1, 3) * B(3, 1) +
          A(1, 4) * B(4, 1)

C(1, 2) = A(1, 1) * B(1, 2) + A(1, 2) *
          B(2, 2) + A(1, 3) * B(3, 2) +
          A(1, 4) * B(4, 2)

C(1, 3) = A(1, 1) * B(1, 3) + A(1, 2) *
          B(2, 3) + A(1, 3) * B(3, 3) +
          A(1, 4) * B(4, 3)

C(1, 4) = A(1, 1) * B(1, 4) + A(1, 2) *
          B(2, 4) + A(1, 3) * B(3, 4) +
```

```
            A(1, 4) * B(4, 4)

'I = 2
C(2, 1) = A(2, 1) * B(1, 1) + A(2, 2) *
          B(2, 1) + A(2, 3) * B(3, 1) +
          A(2, 4) * B(4, 1)

C(2, 2) = A(2, 1) * B(1, 2) + A(2, 2) *
          B(2, 2) + A(2, 3) * B(3, 2) +
          A(2, 4) * B(4, 2)

C(2, 3) = A(2, 1) * B(1, 3) + A(2, 2) *
          B(2, 3) + A(2, 3) * B(3, 3) +
          A(2, 4) * B(4, 3)

C(2, 4) = A(2, 1) * B(1, 4) + A(2, 2) *
          B(2, 4) + A(2, 3) * B(3, 4) +
          A(2, 4) * B(4, 4)

'I=3
C(3, 1) = A(3, 1) * B(1, 1) + A(3, 2) *
          B(2, 1) + A(3, 3) * B(3, 1) +
          A(3, 4) * B(4, 1)

C(3, 2) = A(3, 1) * B(1, 2) + A(3, 2) *
          B(2, 2) + A(3, 3) * B(3, 2) +
          A(3, 4) * B(4, 2)

C(3, 3) = A(3, 1) * B(1, 3) + A(3, 2) *
          B(2, 3) + A(3, 3) * B(3, 3) +
          A(3, 4) * B(4, 3)

C(3, 4) = A(3, 1) * B(1, 4) + A(3, 2) *
          B(2, 4) + A(3, 3) * B(3, 4) +
          A(3, 4) * B(4, 4)

'I=4
C(4, 1) = A(4, 1) * B(1, 1) + A(4, 2) *
          B(2, 1) + A(4, 3) * B(3, 1) +
          A(4, 4) * B(4, 1)

C(4, 2) = A(4, 1) * B(1, 2) + A(4, 2) *
          B(2, 2) + A(4, 3) * B(3, 2) +
          A(4, 4) * B(4, 2)

C(4, 3) = A(4, 1) * B(1, 3) + A(4, 2) *
```

```
              B(2, 3) + A(4, 3) * B(3, 3) +
              A(4, 4) * B(4, 3)

C(4, 4)  =  A(4, 1) * B(1, 4) + A(4, 2) *
              B(2, 4) + A(4, 3) * B(3, 4) +
              A(4, 4) * B(4, 4)
```

Listing 7.7 **Simpson's Rule integration using rolled-up loops**

```
h_3 = (b-a)/3N
s0=g(0) + g(N)
s1=0
s2=0

For J = 1 to N-1 Step 2

       s1 = s1 + g(J)
       s2 = s2 + g(J+1)

Next J

s1 = s1*2
s2 = s2 * 4

I = h_3 * (s0 + s1 + s2)
```

Listing 7.8 **Simpson's Rule integration loop partially unrolled**

```
For J = 1 to 9993 Step 4

       s1 = s1 + g(J) + g(J+2)
       s2 = s2 + g(J+1) + g(J+3)

Next J

'clean up extra terms
s1 = s1 + g(9997)
s2 = s2 + g(9998)
```

Listing 7.9 Simpson's Rule integration loop further unrolled

```
For J = 1 to 9997 Step 6

      s1 = s1 + g(J) + g(J+2) + g(J+4)
      s2 = s2 + g(J+1) + g(J+3) g(J+5)

Next J

'clean up extra terms

s1 = s1 + g(9997)
s2 = s1 + g(9998)
```

Listing 7.10 General Partially Unrolled Simpson's Rule Loop

```
LoopBreak = (int(N-2)/8) * 8) + 1

LoopEnd = LoopBreak - STEP

'main summation loop
For J = 1 to LoopEnd Step 8

      s1 = s1 + g(J) + g(J+2) + g(J+4) + g(J+6)
      s2 = s2 + g(J+1) + g(J+3) g(J+5) + g(j+7)

Next J

'clean-up extra terms
For J = LoopBreak to N-2 Step 2

      s1 = s1 + g(J)
      s2 = s2 + g(J+1)

Next J

'continue with algorithm
```

Listing 8.1 Basic Scientist Style LJ Pair Energy

```
' Depth is well depth
' SIGMA is related to well position
```

```
{Depth, Num_Particles and SIGMA set in code}

LJ_Energy = 0

For I = 1 to Num_Particles - 1

   For J = I + 1 To Num_Particles

      r = sqr((x(I)-x(J))^2 + (y(I)- y(J))^2)

      LJ_Energy = LJ_Energy + [ (SIGMA/r)^12 -
      (SIGMA/r)^6 ]

   Next J

Next I

LJ_Energy = LJ_Energy * C
```

Listing 8.2 ‘ Scientist CISC style coding of simplified LJ Energy

```
' just a value to use
r = 1.634

' Scientist Style enters
' equation as written on paper
Energy = 1/r^12 - 1/r^6
```

Listing 8.3 ‘ a RISC style version of Listing 8.2

```
' just a value
r = 1.634


'r6 = r⁶
r6 = r * r * r* r * r * r

'r12 = (r⁶)² = r¹²
r12 = r6 * r6
```

```
'Energy = 1/r12 - 1/r6
Energy = 1/r12 - 1/r6
```

Listing 8.4 **' a RISC style variant of Listing 8.3**

```
' just a value to use
r = 1.634

'r6 = r3
r6 = r * r * r

'r6 = (r3)2 = r6
r6 = r6 * r6

'r12 = (r6)2 = r12
r12 = r6 * r6

'Energy = 1/12 - 1/r6
Energy = 1/r12 - 1/r6
```

Listing 8.7 **Scientist CISC style for computation of 2-d particle distance**

```
' some made-up values
' for the two coordinates
Const x1 As Double = 1.234
Const x2 As Double = 2.675
Const y1 As Double = 0.045
Const y2 As Double = 2.073

' 10,000,000 computations
' common in Monte Carlo procedures
For counter = 1 To 10000000

      ' Scientist Style: equation as written
      r = Sqr((x1 - x2) ^ 2 + (y1 - y2) ^ 2)

Next counter
```

Listing 8.8 RISC Style for computation of 2-d particle distance

```
' some made-up values for the
' two coordinates.
Const x1 As Double = 1.234
Const x2 As Double = 2.675
Const y1 As Double = 0.045
Const y2 As Double = 2.073

Dim TempX As Double
Dim TempY As Double

For counter = 1 To 10000000

        TempX = x1 - x2
        TempX = TempX * TempX
        TempY = y1 - y2
        TempY = TempY * TempY
        TempY = TempY + TempX

        r = Sqr(TempY)

Next counter
```

Listing 8.9 a Scientist Style Radial Distribution Function calculation

```
Within_Distance = 0
For Particle_Counter = 1 To Number_of_Particles


    ' compute distance between current particle
    ' and reference particle

    tempX = X_reference - X(Particle_Counter)
    tempX = tempX * tempX

    tempY = Y_reference - Y(Particle_Counter)
    tempY = tempY * tempY

    tempY = tempX + tempY

    Distance = sqr(tempY)

    ' count particle if within interesting
    ' distance
    If Distance <= 20 Then
```

```
        Within_Distance = Within_Distance + 1
    End If

Next Particle_Counter
```

Listing 8.10 Simplified Radial Distribution Function Calculation

```
Within_Distance = 0
For Particle_Counter = 1 To Number_of_Particles


    ' compute distance between current
    ' particle and reference particle

    tempX = X_reference - X(Particle_Counter)
    tempX = tempX * tempX

    tempY = Y_reference - Y(Particle_Counter)
    tempY = tempY * tempY

    ' this is Distance²
    Distance2 = tempX + tempY

    ' compares Distance² to 20²
    If Distance2 <= 400 Then
        Within_Distance = Within_Distance + 1
    End If

Next Within_Distance
```

Listing 8.11 Combined r and LJ Energy calculation with no sqr() function

```
TempX = x1 - x2
TempX = TempX * TempX
TempY = y1 - y2
TempY = TempY * TempY
TempY = TempY + TempX

'don't need to compute sqr(r) then re-square
'it in the energy calculation.  r^2 is used
'to compute r^6 directly as r^6=(r^2)^3

r6 = TempY * TempY * TempY
r12 = r6 * r6
```

```
Energy = 1 / r12 - 1 / r6
```

Listing 9.1 Explicit Calculation of pair energy each time through the loop

```
'r is selected in code

' Compute Energy using RISC style from
' Chapter 8

r6 = r * r * r * r * r * r
r12 = r6 * r6

ThisEnergy = 1 / r12 - 1 / r6
```

Listing 9.2 Using look-up table, precomputed and stored in array En(r)

```
' the array En(r) is precomputed

' for r=0 to 2829.0 pm

'r is selected in code; r is a double

Index = CLng(r * 10)

ThisEnergy = En(Index)
```

Listing 9.3 Combined r and LJ Energy using LJ Energy Look-Up Table

```
'En(r) is array initialized to
' 1/r^12 - 1/r^6

' some made-up values for the
' two coordinates
Const x1 As Double = 1.234
Const x2 As Double = 2.675
Const y1 As Double = 0.045
Const y2 As Double = 2.073
Dim rIndex as Long
Dim TempX As Double
Dim TempY As Double

TempX = x1 - x2
TempX = TempX * TempX
```

```
TempY = y1 - y2
TempY = TempY * TempY
TempY = TempY + TempX

r = Sqr(TempY)
rIndex = CLng(r * 100)

ThisEnergy  = En(rIndex)
```

Listing 9.4 Initialize LUT for r as two dimensional array

```
' The array itself is RVal and is declared as LONG so type
' conversions are not
' done inside the iterative loop (that is, when looking up '
Energy(RVal)).


For deltaXCounter = 1 to 2829

     TempX = CDbl(deltaXCounter)
     TempX = TempX * TempX

     For deltaYCounter = 1 to 2829


          TempY = CDbl(deltaYCounter)
          TempY = TempY * TempY

          RTemp = Sqr(TempX + TempY)

          RVal(deltaXCounter, deltaYCounter)
                                  = CLng(RTemp)

     Next deltaYCounter

Next deltaXCounter
```

Listing 9.5 Actual LJ Energy Function using LUT for both r and Energy

```
Function Homog_LJ_Energy() as Double

     ' Declarations go here
     ' recall that Energy and RVal are
     ' globally declared arrays X and Y are
     ' global arrays holding x,y
     ' position of particle i
```

```
    ThisEnergy = 0

       I is a long, and N is number of
  particles
    For I = 1 to N-1

         For J= I+1 to N        ' J is a long

              deltaX = CLng(ABS(X(I) - X(J)))
              deltaY = CLng(ABS(Y(I) - Y(J)))

              R = RVal(deltaX, deltaY)

              ThisEnergy = ThisEnergy
                              + Energy(R)

         Next J

    Next I

    Homog_LJ_Energy = ThisEnergy

End Function
```

Listing 9.6 Initialization of Heterogeneous LJ Energy Look Up Table

```
' storing energy table with multiple
' interactions declarations go here

' Depth(a) is array holding three interaction
' 'strengths'
' Sigma(a) is array holding three interactions
' 'separations'

For RCounter = 1 to 5657

    R = CDbl(RCounter/100)

    R6 = R * R * R * R * R * R
    R12 = R6 * R6

    Sigma6 = Sigma(1) * Sigma(1) * Sigma(1) *
                Sigma(1) * Sigma(1) * Sigma(1)
    Sigma12 = Sigma6 * Sigma6
```

```
      E(1, RCounter) = Depth(1) * (Sigma12/R12 -
                       Sigma6/R6)

      Sigma6 = Sigma(2) * Sigma(2) * Sigma(2) *
               Sigma(2) * Sigma(2) * Sigma(2)
      Sigma12 = Sigma6 * Sigma6

      E(2, RCounter) = Depth(2) * (Sigma12/R12 -
                       Sigma6/R6)

      Sigma6 = Sigma(3) * Sigma(3) * Sigma(3) *
               Sigma(3) * Sigma(3) * Sigma(3)
      Sigma12 = Sigma6 * Sigma6

      E(3, RCounter) = Depth(3) * (Sigma12/R12 -
                       Sigma6/R6)

Next RCounter
```

Listing 10.1 Computing only energy change in Single Particle LJ Energy routine

```
Function Single_Particle_LJ (MovedIndex as Long,
          TotalNum as Long, XMoved as  Long,
          YMoved as Long) as Double

' Declarations here
' recall Energy and Distance are globally
' accessible arrays

ThisEnergy = 0
For ParticleCounter = 1 to N

      rx=particles(ParticleCounter).x
      ry=particles(ParticleCounter).y

If ParticleCounter <> MovedIndex Then

      deltaX = ABS (CLng(rx - XMoved))
      deltaY = ABS (CLng(ry - YMoved))
      R = RVal (deltaX, deltaY)
      ThisEnergy = ThisEnergy + Energy(R)

End If

Next ParticleCounter
```

```
Single_Particle_LJ = ThisEnergy

End Function
```

Listing 10.2 Sample User Defined Data Type for Heterogeneous (multi-particle) LJ Energy

```
'This goes in declaration section
Type Particle

     Type As Boolean
     X(2829) as Double
     Y(2829) as Double

End Type

Dim Particles(NumberOfParticles) as Particle
```

Listing 10.3 Heterogeneous LJ Energy via If...Then structure to select interaction type

```
' Note: Test Expression is evaluated multiple
' times
' access the Ith and Jth particle types via
' Particles(I or J).Type

If Particles(I).Type = True AND
          Particles(J).Type = True Then

     EnergyType = 1

ElseIf  Particles(I).Type = False AND
          Particles(J).Type = False Then

     EnergyType = 2

ElseIf Particles(I).Type = True AND
          Particles(J).Type = False Then

     EnergyType = 3

ElseIf Particles(I).Type = False AND
```

```
        Particles(J).Type = True Then
    EnergyType = 3
End If
```

Bibliography

Books:

Advanced Micro Devices, *3DNow! Technology Manual* (AMD, 2000)

Advanced Micro Devices, *AMD Athlon Processor x86 Code Optimization Guide* (AMD, 2002)

Aris, Rutherford, *Vectors, Tensors and the Basic Equations of Fluid Mechanics* (Dover: New York, 1962)

Atkins, P.W. *Physical Chemistry, Third Edition* (Freeman: New York, 1985)

Casad, Joe *SAMS Teach Yourself TCP/IP in 24 Hours, Second Edition* (SAMS: Indianapolis, 2001)

Chandler, David *Introduction to Modern Statistical Mechanics* (Oxford: New York, 1987)

Christianson, Tom and Torkington, Nathan *Perl Cookbook* (O'Reilly: Cambridge, 1998)

de Groot, S.R. and Mazur, P. *Non-Equilibrium Thermodynamics* (Dover: New York, 1984)

Dettman, John W. *Applied Complex Variables* (Dover: New York, 1965)

Foster, Ian "Designing and Building Parallel Programs," www-unix.mcs.anl.gov/dbpp, 1995

Gookin, Dan *Advanced MS Dos Batch File Programming* (Windcrest: Blue Ridge Summit, 1989)

Hill, Terrell L. *An Introduction to Statistical Thermodynamics* (Dover: New York, 1960)

Kreyszig, Erwin *Advanced Engineering Mathematics, Sixth Edition* (Wiley and Sons: New York, 1988)

LaMothe, Andre, *Tricks of the Windows Game Programming Gurus* (SAMS: Indianapolis, 1999)

Mader, Charles L., *Numerical Modeling of Explosives and Propellants, Second Edition* (CRC Press: Boca Raton, 1998)

Mason, W.H. *Applied Computational Aerodynamics,* http://www.aoe.vt.edu/~mason/Mason_f/CAtxtTop.html

McQuarrie, Donald A. *Statistical Thermodynamics* (University Science Books: Mill Valley, 1973)

Microsoft, *Microsoft Macro Assembler 5.1 Programmers Guide* (Microsoft, 1987)
Microsoft, *Microsoft Visual Basic 6.0 Programmer's Guide* (Microsoft Press, 1998)
Microsoft, *Microsoft Visual C++ 6.0 Programmer's Guide* (Microsoft Press, 1998)
Microsoft, *Microsoft Windows 2000 Professional Resource Kit* (Microsoft Press, 2000)
Morse, Stephen P. and Albert, Douglas J., *The 80286 Architecture* (Wiley and Sons: New York, 1986)
Mueller, John Paul, *Poor Richard's Home and Small Office Networking* (Top Floor: Lakewood, 2001)
Oliver, Dick *SAMS Teach Yourself HTML 4 in 24 Hours, Fourth Edition* (SAMS: Indianapolis, 1999)
Schildt, Herbert, *C: The Complete Reference, Second Edition* (Osbourne McGraw-Hill: Berkeley, 1987)
Spector, David M., *Building Linux Clusters* (O'Reilly: Cambridge, 2000)
Versteeg, H.K. and Malalasekera, W. *An Introduction to Computational Fluid Dynamics: The Finite Volume Method* (Prentice Hall: London, 1995)
Wangsness, Roald K., *Electromagnetic Fields, Second Edition* (John Wiley and Sons: New York, 1986)
Wille, Christophe and Koller, Christian *SAMS Teach Yourself Active Server Pages in 24 Hours* (SAMS: Indianapolis, 1999)
Zaks, Rodnay, *How to Program the Z-80, Third Edition* (SYBEX, 1982)

Articles:

DeTar, Carlton and Gottlieb, Steven "Lattice Quantum Chromodynamics Comes of Age," *Physics Today* 57 (2004) 45.
Fisher, Jason "Take Charge with Windows Management Instrumentation," Microsoft Developers Network, 2000
Fletcher, Drew "Optimizing Microsoft Visual Basic 4.0" Microsoft Developer's Network, 1995
Jayasimha, D.N., Hayder, M.E. and Pillay, S.K. "Parallelizing Navier-Stokes Computations on a Variety of Architectural Platforms," *ACM* (1995)
Lammers, Don "Connecting HTML Help to Visual Basic Programs," www.smountain.com/resource/VBHTMLHelp.pdf
Mathisen, Terje "Pentium Secrets," on LaMothe, Andre, *Tricks of the Windows Game Programming Gurus* (SAMS: Indianapolis, 1999) CD-Rom, 1996
Norris, SE and Armfield, S.W. "Solving the Navier-Stokes Equations on a Workstation Cluster," *Anziam J.* 42E (2000) C1058
Pietrek, Matt "A Programmer's Perspective on New System DLL Features in Windows NT 5.0, Part I," Microsoft Developer's Network, 2002
Rogerson, Dale "Microsoft Windows and the C Compiler Options," Microsoft Developer's Network, 1992

Sarma, Debabrata "DLLs for Beginners," Microsoft Developer's Network, 1998
Vasudevan, Alavoor "PHP How-To," ftp://sunsite.unc.edu/pub/Linux/docs/HOWTO/PHP-HOWTO

Web Sites and Pages:

allserv.rug.ac.be/~tkuppens/chem.shtml

This is Tom's Free Chemistry Software Page. This site contains alphabetized and searchable lists of free software from molecule drawing packages, utilities and various modeling packages.

developer.intel.com/software/products/vtune/index.htm

Commercial Intel software profilers.

ece.Clemson.edu/parl/grendel.htm

A description of the 18 node Grendel cluster, part of Clemson University's Parallel Architecture Research Laboratory.

en.wikipedia.org/wiki/Motorola_68000

A summary of the 68000 family of processors.

en.wikipedia.org/wiki/Motorola_6809

A summary of the 6809 processor and its history.

en.wikipedia.org/wiki/G4

A summary of the MPC7400 family of processors.

foldingathome.standford.edu

This is the Folding@Home site for protein folding distributed computing via the Internet

gcc.gnu.org

The gcc home page.

gah.standford.edu

Genome@Home site for gene encoding distributed computing via that Internet.

math.jpl.nasa.gov/nr/

This is the page titled "Why not use Numerical Recipes." The first line in the page is "We have found *Numerical Recipes* to be generally unreliable." The page is based mostly on older editions of the book.

msdn.microsoft.com/vstudio/techinfo/articles/upgrade/Csharpintro.asp

This is an introduction to MS C#.

osdev.neopages.net/tutorials/gccasmtut.php

An introduction to inline ASSEMBLY programming with gcc.

setiathome.ssl.berkeley.edu

The Seti@Home distributed computing project for analyzing radio astronomy signals in the Search for Extraterrestrial Intelligence.

support.microsoft.com/?kbid=216181

This is Microsoft Knowledge Base Article 216181 regarding production of incorrect code that results when using the /Og compiler optimization.

support.microsoft.com/?kbid=234602

This is a Microsoft Knowledge Base Article 234602 regarding production of incorrect code that results when using the /Og compiler optimization.

tech-report.com

Reviews of hardware, such as cpu tests and benchmarks.

http://www-106.ibm.com/developerworks/linux/library/l-rt9/?t=gr,lnxw01=ConSwiP1

Ed Bradford's article comparing context switching in Linux and Windows.

www-106.ibm.com/developerworks/linux/library/l-rt7/?Open&t=grl,l=252,p=mgth

Ed Bradford's article comparing process and thread creation in Linux and Windows.

www.aceshardware.com

Reviews of hardware, such as cpu tests and benchmarks.

www.amdzone.com

In particular, the page at

www.amdzone.com/articleview.cfm?articleid=282

has a tutorial for MMX and 3DNow! programming.

www.arstechnica.com/

The Ars Technica PC Enthusiast Resource online, including articles about hardware and theory. Also included are articles on system performance tuning.

Specifically, the article by Jon Stokes at

www.arstechnica.com/cpu/lq00/g4vsk7/g4vsk7-1.html

and subsequent pages contain a detailed comparison of the G4 to the K7 processors.

Also, Jon Stokes' article beginning at

http://www.arstechnica.com/paedia/r/ram_guide/ram_guide.part3-1.html

and subsequent pages contain detailed descriptions of microcomputer memory subsystems.

www.aspemporium.com

Repository of ASP related tools and discussions. In particular, the page at

www.aspemporium.com/aspEmporium/tutorials/GetObject/WinMgmts_moniker.asp

includes an overview of the WMI moniker WinMgmts. This page is included on the CD-Rom by permission of the author.

www.barefeats.com

Reviews of hardware, such as cpu tests and benchmarks.

www.cacr.Caltech.edu/Beowulf/tutorial/building.html

CACR's brief overview of building a Beowulf cluster.

www.cfxweb.net

Reviews of hardware, such as cpu's. In particular, the page at

www.cfxweb.net/articles/hugi21/pcp4.shtml

discusses the Pentium 4 processor in technical detail.

Also, the page at

www.cfxweb.net/articles/hugi21/cosimdfp.shtml

has an article about 3DNow! SIMD optimization.

www.chem.ox.ac.uk/curecancer.html

Screensaver Lifesaver distributed computing project for researching interactions between cancers and drugs.

www.codeplay.com/manual

Online manual for the VectorC compiler.

www.cortstratton.org/content/tutorials/HugiCode.html

Evolutionary development of vector/matrix optimization using the SSE instruction set

www.cpuid.com

CPU instruction set summaries

www.epm.ornl.gov/pvm/intro.html

An introduction to programming with PVM.

www.extremetech.com

Reviews of hardware, such as cpu tests and benchmarks.

www.flipcode.com/tutorials/tut_perform.htm

A tutorial on code optimization.

www.freeos.com/articles/3185/

This page summarizes important command line options for gcc.

www.freshwater.com/support/notes/noteTN10333.htm

Tips for enabling remote connections for remote WMI in Windows 98.

www.ganssle.com/articles/abuscyc.htm

Though dealing with old (Z-80) technology, this page provides a good basic discussion of processor bus timing and cpu cycles.

www.jorgon.freeserve.co.uk/TestbugHelp/MMXins.htm

An introduction to the MMX instruction set.

www.linux-tutorial.info

This site is an overview of Linux from a system administrator perspective, including a very good comparison of Linux to Windows (the author is an experienced WinNT administrator).

www.linuxcommand.org

This site is a beginner tutorial for using Linux. Linux files system structure is outlined and interacting with the system via the Shell is presented. In addition, numerous shell scripts are provided.

www.linuxgazette.com

This is an online 'e-zine' for linux.

www.linuxplanet.com/linuxplanet

This site has forums for Linux in Enterprise environments.

www.mersenne.org

GIMPS distributed Mersenne Prime Number Search.

www.mvps.org/htmlhelpcenter/htmlhelp/hhvbclas.html

HTML Help Class for Visual Basic

www.netlib.org/linalg/html_templates/node20.html

This site has a brief description of the Conjugate Gradient method for linear system solution.

www.pcguide.com

In particular, the pages at

www.pcguide.com/ref/cpu/arch/int/execNative-c.html www.pcguide.com/ref/cpu/arch/int/execTranslation-c.html

www.pcguide.com/ref/cpu/arch/int/featSpeculative-c.html

give descriptions of cpu instruction execution cycles, Speculative Execution and Branch Prediction

www.penguin.cz/~literakl/intel/intel.html

An overview of the x86 instruction set.

www.platformdev.com/

Commercial Visual Basic code profilers.

www.quantasm.com/index.html

x86 and x87 basic instruction set latencies and terminology glossary.

www.sara.nl/Beowulf/

An introduction to the Beowulf cluster at SARA.

www.scl.ameslab.gov/Projects/ClusterCookBook/intro.html

Brief overview of building a basic, four node cluster.

www.stereopsis.com/FPU.html

This page is an overview of floating point bottlenecks caused by poor program construction. Of particular interest are the sections on float to integer conversions and precision control.

www.tom.womack.net/x86FAQ/faq_cores.html

An overview and technical description of x86 cpu evolution.

www.tweak3d.net

Hardware and OS performance tuning.

www.winguides.com/registry

Windows Registry Guide; lists many registry keys and explanations, with examples.

www.zdnet.com

Reviews of hardware, such as cpu tests and benchmarks.

vergil.chemistry.gatech.edu/resources/programming/c-tutorial/libraries.html

This page outlines the process of how to create libraries in gcc.

Index